DATE DUE			

THE CHIMPS OF MT. ASSERIK

THE CHIMPS OF MT. ASSERIK

STELLA BREWER

ALFRED A. KNOPF NEW YORK 1978

THIS IS A BORZOI BOOK
PUBLISHED BY ALFRED A. KNOPF, INC.

Library of Congress Cataloging in Publication Data

Brewer, Stella Margaret [date]
The chimps of Mt. Asserik.

Includes index.
1. Chimpanzees. 2. Mammals—Senegal—Assirik, Mount. I. Title
QL737.P96B73 1978 599'.884 77-20355
ISBN 0-394-49449-0

The author wishes to thank Edward Brewer, Nigel Orbell, Raphaella Savinelli, Michael Shaw and Hugo van Lawick for permission to use some of the photographs in this book.

Manufactured in the United States of America

FIRST AMERICAN EDITION

To My Mother and Father

CONTENTS

PART THREE: LEARNING TO BE FREE

MAPS

ACKNOWLEDGMENTS

IT WOULD HAVE BEEN IMPOSSIBLE to set up the rehabilitation project, still less write this book, without the help and encouragement of many people to whom I want to express very sincere thanks:

First of all must come my family, my father especially, for all that they have contributed, and for their constant, unwavering support.

The Gambian government, in particular Mr. Harry Lloyd Evans, acted with exemplary speed and efficiency in ending the trafficking in chimpanzees, and in helping in the rehabilitation of the victims.

Mr. A. R. Dupuy, Director of National Parks in Senegal, and Mr. M. Gueye, Conservator of Niokolo Koba National Park, made the whole project possible by letting me set up camp at Mount Asserik and helping me over all the difficulties. Jane Goodall gave me invaluable personal advice, much help and encouragement, and the opportunity to visit her research center in Tanzania. Miss Molly Badham and her partner, Miss Natalie Evans, the Director of Twycross Zoo, gave me their assistance; their faith in my work meant a great deal to me.

Michael Brambell, Curator of Mammals at the London Zoological Gardens, has supported the idea from the first, and I can never thank him enough for the crucial part he played in sending Yula and Cameron to Africa to join the project.

Ted Scrope Howe devoted much hard work and time to fund-raising for the project, and to making the arrangements for Yula and Cameron's flight to West Africa.

Without Claude Lucazan's generous help life in camp would have been very much more difficult, perhaps impossible, and my

Land Rover would certainly not still be navigating the road from The Gambia to Mount Asserik.

Nigel Orbell's friendship over the years and his help in looking after the chimps, both in Abuko and Niokolo, have made a unique contribution to the whole enterprise.

Raphaella Savinelli, my partner, brought Bobo to camp and has put much hard work into the project. Her companionship has brightened an otherwise rather lonely life in the bush.

Réné and Julian have showed great loyalty and dedication to the chimps. I vastly appreciated their cheerful company throughout the first two years of the project, when there were only the three of us in camp.

Abduli and Alhadji helped to look after the chimps at Abuko, as did the other members of the staff—Bojang and Doodoo especially—who struggled through the early years to make Abuko what it is today.

Grannie, Edna, Fleure and Pusska took me into their home in Ilford, and looked after me for four months while I finished writing this book. Jan Morrison also gave me hospitality and kindly typed the first part of the manuscript.

The Fauna Preservation Society gave valuable support; also Sam and Sue Samaraweera. Bruce Wolfe and many other individuals have made donations to the project. Without so much generous support we could not have continued.

Hugo van Lawick from the first has shown great enthusiasm for the project and has helped me generously in many ways to achieve my goal. I am certain that there are few filmmakers who would have shown such patience and consideration while filming the chimps, or who could have achieved a more accurate representation of life in camp.

Finally, the late Sir William Collins gave great help and encouragement in setting up the project, and Philip Ziegler showed patience and understanding while waiting for the first draft, and devoted much time and hard work to helping me put the book together.

FOREWORD

NEXT TO MAN HIMSELF, the chimpanzee is surely the most fascinating creature in the world. If we look up "chimpanzee" in a dictionary we find him described as an African ape resembling man. Just how closely he resembles man was not known until naturalists began to watch his behavior in the mountains and forests of his homeland. In the Gombe National Park, in Tanzania, I have been watching chimpanzees for eighteen years, following the life histories of known individuals and learning, all the time, more and more about their fascinatingly complex social behavior.

Chimpanzees are found in the equatorial belt of tropical Africa. They are adapted to a life in the forest and spend much time up in the trees, feeding on luscious fruits and shoots where it is cool and they can catch every breeze. They sleep in the trees, on leafy platforms. In moments of excitement they swing wildly through the branches, swaying great boughs, filling the air with their clamor. At least, this is the kind of life which most chimpanzees lead, and which all would lead except for human interference.

Unfortunately for the chimpanzees, man evolved along with his ape cousins, in the process developing his extraordinary ability to change the nature of things, his desire for knowledge and entertainment, his terrible capacity for destruction. Man has cut down forests in order to cultivate food and build houses, thus destroying much land which used to be home for the chimpanzee. He has shot chimpanzees to feast on their flesh. He has captured baby chimpanzees and transported them to far-off lands to learn about them, gape at them, laugh at them and sometimes to love them—but always in his own environment. For man, having been given authority over the fishes of the sea, the fowls of the air and the beasts of the land, can do with them—or so he often seems to think—precisely as he wishes.

Fortunately there are those with greater understanding, less conceit and a more perceptive scale of values. Several years ago I received a letter from a Mr. Brewer in The Gambia. He told me some interesting things about the orphan chimpanzees which his

A mother chimp and her baby photographed at Jane Goodall's Gombe Research Center, Tanzania

daughter was caring for (and watching) in their Nature Reserve at Abuko. That was when I first heard about Stella. We corresponded for a while, and I soon felt that I knew, personally, William, Tina, Pooh and the rest. I was impressed by Stella's obvious dedication to wildlife, and also her ability to make careful observations. Finally we met in London.

Her ambition was to reintroduce into the forests of Senegal chimpanzees which had lived for years in captivity or which had even been born there. It was a challenging project, for it would not be enough merely to release the apes into suitable country: they would have to be trained to cope with life in the wild; be taught what dangers threatened and how to meet them, what foods they could eat and where to find them, how to crack nuts, fish for ants, make nests. If Stella was to teach her chimps how to survive, she must first watch wild chimpanzees living their natural life. I invited her to Gombe, where, as she tells in this book, she was able to study wild chimpanzees at close quarters and find out how *her* chimpanzees ought to live when the moment arrived for their release.

Since those days I have followed with admiration and keen interest the story of Stella's attempt to rehabilitate her orphans. If anyone has the patience, the dedication and the courage to make the experiment work, it is she. Already she has achieved far more success than many of us thought possible. At the same time, she and her chimpanzees are making a valuable contribution to our understanding of these most human of all the apes: how they adapt to a new life, with new foods, new dangers, new values.

Often, as I watch the free chimpanzees at Gombe, I think about the sad prisoners in other places: chimpanzees behind bars, in solitary confinement, undergoing painful experiments, performing stupid tricks, sitting, day after day, year after year, with nothing to look forward to. It is good to know that, if Stella's goal is achieved, some of them might have a chance to return to the forest which is their birthright.

Jane Goodall

PART ONE

BEGINNINGS

1. WILLIAM

I LOOKED UP AND WATCHED HIM FEED. He sat, engrossed in his meal and quiet for the time being. He stretched out a long arm and drew a cluster of violet mandico berries to him. His mobile lips neatly removed one berry after another, until his mouth was full. He leaned against the broad trunk behind him and settled down contentedly to enjoy them. Idly, his intelligent brown eyes watched the quivering cluster spring back to position on its release, and followed a few fruits dislodged by the movement as they sprinkled down on the dry leaves below.

He lifted his hand and raked flexed fingers through the long hair on his upper arm, in an audible and satisfying scratch. Then, as he turned his head, his expression became one of intense concentration. He had spotted a minute speck of dry skin amid the now disheveled hair. Leaning his chin on his upper arm, he pushed his lower lip forward and delicately disentangled a fleck of skin. With it still perched on the tip of his protruding lip, he focused down his nose for a second or two; then his mouth compressed around it, as though to confirm by feel what he had just seen. Again his lip came forward and a second brief scrutiny of the speck of skin took place. Finally satisfied, he drew it in so that it joined the wad of well-sucked berries and was ingested. An industrious search began for other hidden foreign bodies. Diligently his fingers lifted one tiny tuft of hair after another as he groomed his arm, but he found nothing more to alarm him. He relaxed again, to suck on his soggy ball of berries.

Suddenly he moved, leaf-filtered morning sun glancing off his glossy black coat, paling it in patches, illuminating it in others to reflect a silky metal-blue sheen. Stamping and slapping the branches in sheer high spirits, he rushed on to join his companions in their playground among the trees.

I could not help feeling proud at the sight of him. He had come such a long way since we had first met. He was almost seven years old now, strong, healthy and, I believed, happy. Yet there had been a time when we had wondered whether William could survive. As a tiny infant he had been snatched from the life he was born into, robbed of a warm belly to cling to, a nipple to satisfy and soothe him, thick gentle fingers to tickle and groom him and long arms to protect and comfort him: in fact all the things that make up a mother for a little chimpanzee.

I VIVIDLY REMEMBER the day William arrived and the man who stood outside the office, a grimy box at his feet. His tribal scars suggested that he was not a Gambian. As my father and I approached, he bent and began to untie the dirty pieces of string and rag that bound the box beside him. A nauseating odor wafted up as he lifted the lid, a warning of what was to come. Rough hands pulled a rigid little body from inside and placed it on the concrete. There it lay, quite stiff, bony arms and legs drawn close to its body. Its small, pale face contorted into a fixed and terrified grin. Only its chest moved as it emitted a harsh choking scream over and over again. The sparse, dark hair that covered its emaciated frame was matted with filth and pus that ran from the numerous sores all over its body. Its belly protruded, alarmingly swollen and taut.

William had made the three-week journey from Guinea, where he had been captured, in the 12 × 15-inch prison in which he had arrived. For part of the journey his box had been tied to the top of rolling, jolting African buses. The heat and the violent movement of the buses would have been torture enough, but beside that were the thick clouds of suffocating dust which enshroud any vehicle on the bush roads. William, perched where he was, would have been in the midst of what must have seemed an endless nightmare to the confused and terrified young chimp. It was a miracle that he had survived.

He had been secured in his box by a piece of plastic electrical cord, which was fastened around his groin. This had been threaded through a hole in the back of the box. William was then folded like a rag doll, till his face touched his knees, and the cord was drawn from behind, dragging him into his confining quarters. The parts of his body that protruded were crammed into the few small spaces still available.

He was for sale: a murdered mother and weeks of indescribable misery for the sake of a few shillings. We were overwhelmed with pity for the pathetic creature at our feet. After some hurried bartering, money was placed in the outstretched hand. We little realized then what the consequences would be.

I wrapped the chimp in a sack. He weighed about 6 pounds and looked to be about 15 inches long. He was about 2 years old. Gently cradling the cringing body close to mine, I walked back to our home. A spacious crate was lined with clean packing straw and placed on the verandah. To remove the worst of the filth from his coat, William was wiped with cotton wool, which we had soaked in a mild antiseptic, and his sores were carefully cleaned and treated. We spoonfed him a small amount of baby porridge, liberally sweetened with glucose, most of which he refused. Then we placed him

William, eight weeks after he arrived

on the bed of clean straw. Gratefully he cuddled a bundle of his bedding and fell into an exhausted sleep.

For six weeks he did little else. At first whenever we approached he would huddle against the back of the crate, as far away from us as he could get. Clutching his mat of straw he would whimper or scream. Tess, our little mongrel bitch, was fascinated by William from the beginning. William was extremely timid, but as Tess was of such a gentle disposition, we allowed her to visit him. Whenever she could she would slip out onto the verandah and lie in front of William's crate. The last few yards would be traveled on her stomach, wiggling along, tail wagging slightly, with an expression on her face which combined "eager to please" and "caution." On reaching the entrance, she would place her chestnut head on her front paws and lie patiently for hours, occasionally giving a whimper when William stirred.

As the days passed William's attitude changed from fear to mild irritation. He would shake his hand at her or wave her away with a sweeping upward movement of his thin arm. Gradually, however, Tess's patience was rewarded. As William began to regain his strength, his confidence returned. Tess was the first living creature we saw him approach and touch of his own accord. He had been in the house almost six weeks when one day, toward the end of Tess's usual afternoon vigil, William awoke. He sat up sleepy-eyed, a tuft of wood-wool perched comically above one ear. He gazed at Tess, then cautiously extended spidery fingers and touched her face. Tess raised her head, whimpered and wriggled closer, burying her moist nose in the straw. Her tail thumped out a slow, even rhythm on the tiles.

William had withdrawn his hand the instant Tess moved her head, but shortly afterward the tentative touch was repeated. The beat on the tiles quickened but otherwise Tess remained still. All seemed to be going well when Tess burst forth in a mighty sneeze. William let out a squawk and retreated to the back of his crate. Poor Tess did her best to apologize. Whining and twitching, she attempted an appeasing lick, but William was at least two feet away.

The throb on the tiles had meanwhile grown louder. Slowly it subsided again as Tess regained her composure and once more resigned herself to watching. William began to relax, the grin of fear on his face disappeared, and he pulled gently at the pieces of wood-wool around him. At last he began to cast frequent glances at Tess. Then he tried to incorporate her in his quiet play by flicking pieces of straw in her direction. Tess dared do nothing save whine gently. Slowly he inched his way toward her again, until he was close enough to touch a piece of straw that dangled from her ear. This time Tess managed to avoid sneezing. By the end of the afternoon William had become confident enough to finger her ears and face and was even so bold as to pat her neck once or twice.

Tess behaved perfectly, despite her obvious excitement. From that day on their friendship flourished. William even got over his squeamishness at being licked, and Tess, in a maternal fashion, would try to keep him clean.

He began to come forward eagerly for his small but frequent meals. On hearing our footsteps, a drowsy little face would peek round the edge of the crate, and thin hands were extended gravely to accept our offerings. Slowly he came to accept us as he had Tess. When we left him he would whimper and take a few wobbly steps after us, as if he wanted to follow but was loath to leave his familiar crate. Eventually the day came when security meant one of us and not a bundle of lifeless straw. Once William reached this stage we decided that it was safe for him to make the acquaintance of the other, more boisterous members of our family.

In one sense this story begins with William's arrival into the household, and yet it could not be complete if I did not say something about the background against which it unfolded. Here, then, is a brief account of a Gambian childhood.

2. A GAMBIAN CHILDHOOD

I HAVE LIVED IN THE GAMBIA most of my life. Our home is fifteen miles outside the capital, Banjul, formerly Bathurst, on Yundum Agricultural Station, where my father then held the position of Forest Officer. When we arrived, in 1958, the garden around our bungalow was a black and charred wilderness, for those preparing for our arrival had burned, rather than cut down, the tall elephant grass that had taken it over since our predecessor had left. In a short time, however, this wilderness made way for a colorful garden. The land at the back of the house was used for vegetables and fruit trees, and that at the front and sides for flowers and decorative shrubs.

There were no other children of our own age in the vicinity, but there was so little difference in age between Heather, my younger sister, then aged four, and me, aged six, that we had no real need for other children. Exploring our new home was one long adventure, and in retrospect I see that we enjoyed an almost idyllic childhood. Most mornings after breakfast we were free to roam the country around us and amuse ourselves as we pleased. There was no real danger in this, for big game had long since disappeared from The Gambia. Although leopards were reported in our area, sightings were few and far between. Perhaps the most potentially dangerous creatures were snakes, but normally the only part of a living snake we saw was the last few inches of a tail, as it hurried to put as much distance between it and us as it could. So long as we were reasonably alert, we were as safe as children who live in towns or near roads. To make sure we came to no harm, Satu, our Gambian nanny, kept an eye on us.

As we played, she would point out to us fruits and berries that were edible. In the heat of the day we would sometimes sit in the shade of a large netto tree and Satu would show us how to plait and weave baskets. One by one we discovered the small villages around us. Banjulinding was our favorite, perhaps because it was closer than the others. It was a typical village of mud huts, usually single-roomed rondavels constructed from mud bricks plastered together with yet more mud. Layers of grass thatching made up the roof and these were held to the rhun-palm rafters with bark rope or palm leaves. The influential members of the village had a more Western style of house. These were almost always constructed of corrugated iron, and to me were far less picturesque than their mud fore-runners.

One of our first visitors when we arrived in The Gambia was Momadou, the bagman, and he has been coming regularly to us ever since. Momadou is rotund, jovial and an endearing old rogue. He pedaled out to the stations on a decrepit black bicycle held together ingeniously with bits of string and bark or pieces of wire. Centrally on the handlebars sat a squat wrought-iron bell, which would warn us of his approach when he was still a long way off. On either side of the crosspiece hung two bulging hessian bags, and strapped behind the seat with strips of inner tubing was yet another. Puffing and damp from the strenuous pedaling, he would park his bike at the garden gate and the old bell would sound out loud and clear that Momadou had arrived.

On seeing one of us, he would heave his wares onto his back and struggle down the path with them. Before he left, every single one of the items in the bag had been withdrawn and handed round for inspection. These were mainly wood carvings, local jewelery, and leather artifacts made from animal skins. We would haggle until a price was settled upon, usually a third of the sum Momadou had originally asked for. This bargaining involved much exaggerated acting on both sides, but was an enjoyable and expected part of the purchase.

When everyone concerned had exhausted himself, Momadou would place his objects back in his bags and, with a wide smile, pedal on to the next customers. We usually found each time that we had cunningly been persuaded to buy all sorts of things that we did not have any use for or indeed even like, but we never became immune to his charms. Strange as it may seem now, the fact that some of his wares were made from the skins of wild animals did not then distress us unduly. Perhaps this was because until recent years there was only one Momadou.

My mother and father are both animal lovers. I cannot remember a time when there has not been an animal to share our home with us. I grew up with Pal, my father's golden Labrador. My sister Lorna, who is eight years older than I, had a mania for cats and was always bringing in strays that she had found. In the Seychelle Islands off the east coast of Africa, where I was born, we lived next door to the botanical gardens. I especially liked visiting a large pit that contained some giant tortoises. Sometimes I was allowed to ride them, though I couldn't have been more than three years old at the time. As I struggled to climb aboard, the tortoise would usually stop and draw its head and legs into its shell. Once I was safely perched on its back, Maud, my nanny, showed me how to move my mount. She handed me a small stone and instructed me to rub it backward and forward on the front of the tortoise's shell. It worked; slowly a scaly head would appear, followed hesitantly by four legs, the shell beneath me would rise abruptly and we would be lurching off for a slow but nevertheless exciting walk around the pit.

It was in the botanical gardens that I carried out the first rescue operation that I can remember. Maud and I were walking back to the house when I spotted a small brown bird hopping and flapping about on the path. I carried it home, where I placed it in a cardboard box on the kitchen table and hurried off to call my mother. We returned to the kitchen just in time to see Jackie, one of Lorna's cats, leaving through the back door with the bird in his mouth. I gave chase, screaming at the cat, which only served to make him run faster and I quickly lost him and my little bird. Finally I returned to the house, so upset I could scarcely speak. My mother comforted me, and impressed upon me that I should not blame Jackie for what had happened, for to him eating a bird was a perfectly natural thing to do. It was a lasting introductory lesson on the laws of nature.

When we were transferred to The Gambia, we began to accumulate animals immediately. Ginger came first. He was a beautifully marked marmalade kitten, and to keep him company we adopted Shot, a short-haired German pointer puppy, who was a rich chocolate brown with a white patch on his chest. Shot and Ginger got on very well. As the first of our animals, they viewed the various other species that came to us with an aloof cordiality. They were a perfect pair and, until Shot died four years ago, they demonstrated an unusually deep affection for one another. Ginger, now nineteen years old, is still with us today, and as fit and independent as ever.

Soon after we had arrived, one of our neighbors, who was about to leave The Gambia for good, asked us to care for his two tame birds. One was a yellow-bellied Senegalese parrot, called Mrs. Mop; the other, a long-tailed parakeet called Rolly. We built an aviary for them around one of the large gmelina trees in the garden. Rolly was a sociable, talkative bird and from the verandah one could hear him repeatedly whistling "God Save the Queen" and chatting madly to himself for hours. Mrs. Mop was quite different; she would peck at one on the slightest excuse and never, as far as I can remember, said anything in English, though she would swear aggressively in her native tongue if one persisted in trying to make friends with her.

Shortly after these two had settled in, the rains broke and we found ourselves with twelve more birds. One morning, after a particularly violent storm, Heather and I chanced to pass a large tree laden with weaver bird nests. During the storm many of these intricately woven nests had fallen to the ground. The area beneath the tree was littered with soggy, battered debris. I noticed a lean piebald cat picking his way through the remains, pausing every now and then to inspect a nest. Finally he found what he was looking for. Destroying a nest, he began to feed on its small, naked occupants. I called Heather and we joined the cat in his search for fledglings. Many were already dead, but we found twelve of them still alive. We took off our shirts, which seemed the only dry material at hand, and wrapped the cold bodies in them. Heather suggested that we take some nests too and dry them, so that the young birds would feel at home.

When we got home the nests were duly dried and lined with cotton wool. Under Mummy's guidance we fed the little birds several times a day, placing the food into their ever-gaping mouths with matchsticks. Two of the fledglings died but the remaining ten survived and eventually flew away.

Our interest in animals was noticed and more and more of them, usually orphans, were brought to us. From an early age we were experimenting with different milk formulae to suit each of the species.

Normally rats are killed on sight, but one young farmer who had found a nest of Gambian pouched rats instead brought them as a present to Heather and me. The young rats were pink and blind, and we did not successfully raise them all, but the three that did survive were charming pets. I think they were rather myopic, but their ever-quivering silver whiskers made up for any defect in their sight. They came when they were called and enjoyed being stroked. They didn't object to being picked up and would scramble about all over us. I usually couldn't stand them on my neck for long, as their mobile whiskers tickled unbearably.

When Pinny arrived we were at a bit of a loss to know where to accommodate him. Pinny was an angry young porcupine. My mother had had a pen built in the back garden for the ducks she was raising, so we decided to try Pinny in there until we thought of somewhere else more suitable. We could not have made a better choice. Pinny seemed to fit in remarkably well with the ducks and even better with the ducks' food bowl. The only drawback to this arrangement was that whenever any of us went to feed them, Pinny would chase us out again. I think Mummy was really quite relieved when she discovered that he had escaped one night. He had burrowed his way under the wire and vanished. We never saw him again.

Another pet I particularly remember from that era was Olly, an African barn owl. At first, because of his formidable beak, we fed him with a pair of tweezers, but we soon found that it was safe to use our fingers. He would gently take the food and hold it for a second in his beak. Inclining his head a little, he would blink his huge, dewy eyes as he gulped, and the food would disappear. He seemed to enjoy having his head and neck stroked gently with a finger and would perch on our shoulders for hours each evening. When he could fly confidently, we would take him out into the garden in the evenings and he would circle the house and return to us. Gradually his flights became longer and, finally, the night came when he did not return. A couple of evenings later there was a screeching call from the sill of the sitting-room window. It was Olly, safe and hungry, demanding entrance and some supper. For years Olly visited us. Often weeks would elapse between his visits and we always felt touched and honored when he did reappear, for he was free and obviously self-supporting.

Every Sunday morning my father would take us as a treat to the airport woods, a patch of natural forest close to the airport at Yundum which harbored a rich variety of Gambian animals. We would get up early so as to be in the woods at dawn, then slowly walk through until the sun was well up. The light and the birdsong created a beautiful world on those mornings. Heather and I would grip hands and grin at each other in silent excitement when we came across a troop of western red colobus monkeys, always high up in the trees. They are fairly large monkeys with black backs and chestnut arms and legs and a ropelike tail that hangs down indistinguishable from the mesh of vines dangling from the trees. Sometimes, if we were quiet, we were able to watch them feed, but it was rare that we could enjoy this luxury for long, since they would spot us and flee, making the most spectacular leaps as they went.

Of all the animals with which we were familiar, the monkeys held a special fascination. We would sometimes go out in the mornings specifically to find and watch them. We wove fantasies about our own imaginary troop, and some of the baskets we made with Satu were taken home to our parents as gifts from "the monkeys." We would save scraps from meals and take them out into the bush as payment, in much the same way as some children invent and feed fairies. Then, when I was seven, the fantasy became reality. I was presented with a tiny red patas infant.

Kim, as he was named, was still so young as to be black with contrasting pale face and hands. The joy I felt at having my very own infant to feed and look after was indescribable, and yet even at that selfish age I had room for pangs of regret that Kim had lost his mother. From the time he arrived, we were almost never apart. While he was still relatively helpless, he would cling to the front of my shirt and I would wrap a piece of cloth around my middle to support him. I felt that it was important for him to have something

Stella, age seven, with Kim (left) and Trixie (right)

warm and soft to cling to for the first few months of his life. In this position he would sleep for hours, only waking briefly when he was hungry. My mother would normally prepare his bottle of milk, but I would allow no one else to feed him. It was always difficult leaving him at night, but here my mother was adamant. No matter how much I begged, he was not allowed to sleep with me. I was happy when Trixie came; she was a green vervet monkey and belonged to Heather. Kim and Trixie took an instant liking to each other, and as there were now two of them to share the box at night, it was no longer so distressing to part with Kim.

As soon as our adopted offspring were strong enough, they moved up and would sit on our shoulders, clutching handfuls of our hair to aid their balance. They were lively companions. Kim was a lot more independent than Trixie and yet he was also more protective toward me. If anyone showed the slightest sign of bullying me, a bright rufous bundle of fur would leap to my defense, and would quickly and efficiently distract my opponent by nipping some well-chosen part of his or her anatomy. Not infrequently this would be my mother.

ONE EVENING as Heather and I were scrubbing off the day's dirt, my mother came in to check that we had washed properly. When she had finished inspecting fingernails, she calmly announced that from the next day onward she wanted Heather and me to remain in the house for a couple of hours every morning.

"Whatever for?" we asked in unison.

"We are going to play school," she replied.

At first it was a novelty to have mother teach us simple arithmetic and English, and to try to fill in the pages of the huge workbooks, especially sent out from England for us. However, the novelty soon wore off. Kim and Trixie were, of course, locked outside and would occasionally appear at the window, as if to remind us of all the other, far more exciting things we could be doing. If mother was out of the room and Kim and Trixie appeared persistently enough, Heather and I would look at each other and then jump out of the window to join them. There was trouble when we got back, and once or twice we had to forfeit our visit to the airport woods, but at the time it always seemed worth it.

Finally it was decided that we would both have to go to boarding school in England. Kim and Trixie were to live with a friend of ours at Sapu, a station ninety miles up the river. Here at least they could still run free. Saying good-bye to the monkeys was hard. I had devoted days before in the bush to collecting Kim's favorite foods and spent a long and exasperating time filling a jar with grasshoppers, which were a real delicacy for him. Kim and Trixie sat enclosed in a cage in the back of the Land Rover, obviously discontented with the whole procedure. Heather and I sat with them, feeding them grasshoppers and pieces of fruit through the wire, all the time explaining that we would not be away long. We stayed with them until the Land Rover drove off. We could not conceive it then, but it was the last time we were to see them. In their new home they had no children to spoil them and became wilder and wilder, only coming back to the house for food. Eventually they disappeared altogether.

The next few days were strange. We were all going on leave, so there were crates and boxes, suitcases and trunks, strewn all over the house as our possessions were packed. Heather and I were given a box to pack our toys in, which we filled with all sorts of seemingly important things, ranging from moth-eaten teddy bears to birds' nests. Our toy box must have been repacked one night, for none of our favorite things ever got to England.

Dear old Satu was on the docks to see off her charges. Every now and then she would take off her head-scarf and wipe away the tears.

"Don't cry, Satu," we persisted. "We will be back quite soon, we are only going to school."

Two loud siren blasts later, the M.V. *Apapa* began slowly to heave away from the jetty. We waved until Satu disappeared and The Gambia gradually faded to become no more than a coastline.

3. A HOUSE OF ANIMALS

IT WAS OVER A YEAR before Heather and I returned for the first of the many brief holidays that punctuated the following eight years at school. Each year things were not the same as they had been before. The village grew, and the farms spread farther and farther into the bush behind the compound, destroying favorite places and special trees. Perhaps the saddest thing of all was the gradual disappearance of the airport woods. Each time we came home, more of them had gone to make way for cultivation, for expansion of the airport runway, for the construction of a wider, more sophisticated tarmac road. By the time we left school, the woods no longer existed. Those tall, stately trees with their vine decorations had been felled one by one, and I wondered what had happened to the pack of mongooses, the antelopes, the leaping colobus and all the other creatures whose home the woods had been.

At school Heather and I saw far less of each other than we did at home. All the same, the years I spent there would have been less bearable without her. The letters from home were always full of news about the birds and animals. Daddy, especially, wrote long letters and described any new arrivals so well that we felt we knew them.

Charlie had been given to my father by one of the forest workers. He was a small bundle of soft fur when he came and resembled a spotted Persian kitten with oversized ears. He had been given to Daddy as a *sollo*, which translated means "leopard," but it was obvious from the first that he was not. Later we learned that *sollo n'dingo*, "small leopard," is the name for the serval cat, and that is what the kitten turned out to be.

Two days after Charlie had come, Daddy was given another

Charlie, the serval kitten

spotted infant. Unlike Charlie, this one had a weasely little face and small, bright brown eyes. He had a thick, short coat covered with chocolate-brown spots and blotches, and his long tail was ringed in the same rich color. He was a genet, and my parents called him Tim. The new arrivals, being so young, were sharing the same nursery. Daddy had done all the early bottle-feeding for Tim and Charlie, and they were quite settled by the time we arrived home. Mummy had had her hands full with Bambi and Booful, two harnessed antelope fawns that lived in the back garden. Unlike most of the other animals, they rarely came into the house. They found it difficult to walk on the tiled floors—their hooves would slide on the polished surface and they were in danger of either dislocating or breaking a leg.

Heather and I left school in 1967. It was a glorious feeling of freedom when we realized that we had finished with school forever. The rest of our lives seemed like one long holiday stretching ahead of us.

My return was made still happier by the gift, just before I left England, of a Dalmatian puppy, which I brought with me on the

Tim, the genet

plane. When we reached the house there was happy confusion. Daddy proudly introduced us to Tim the genet and Charlie the serval cat, who were still young enough to accept Heather and me without many qualms. At first they found the puppy a bit over-effusive. They would shy away when the small, pudgy figure, stumbling over its uncoordinated feet and wagging its spotted whip of a tail, rushed up to greet them. Tess, on the other hand, was terribly excited by the new arrival. She followed the pup like a shadow, placing one front paw on it, then the other, licking, nuzzling, and nudging it, inviting it to play. Daddy decided that our new family member was to be called Plum Duff, which inevitably became shortened to Duffy. We could not have chosen a more appropriate name, for in one of the local languages the word *duff* means "stupid" or "silly." Duffy never showed a fraction of Shot's intelligence nor the air of gentle breeding that the old dog possessed. Instead he became the good-natured clown of the family, anxious to be friends and to please all.

After our introduction to Tim and Charlie, Mummy led us into the garden to meet the antelopes, Bambi and Booful. Surprisingly, they came over immediately and began nudging Mummy for a bottle even though Heather and I, two total strangers, stood beside

Tim and Tess at play

her. They were the exact likenesses of Walt Disney's Bambi, with their huge petal-shaped ears and liquid brown eyes. Their legs seemed too long and fragile to support their chestnut bodies. Their sides were patterned with vertical and horizontal stripes, their rumps sprinkled with matching white spots. The complete lack of fear that they showed on that first day was characteristic of both for the rest of their lives. They never panicked as is common to their species whenever anything unexpected occurs; even when adult and free, they were as tranquil as they had been that first day.

Heather and I soon fitted into the routine and relieved our parents of some of the responsibility of caring for the animals. Tim the genet still found the teat of his bottle rather large, but he had nevertheless learned to suck competently. Generally he was far calmer about his feedings than Charlie the serval cat. Charlie had still not learned to control his claws and, in his enthusiasm for the milk, his front paws would reach out and pull the bottle to his mouth. He made no allowances for the hand that fed him, and we had to wear gloves when we gave him his bottle.

Still a baby, Charlie was fond of undignified, kittenish pranks. He would stalk into our bedroom in the mornings and crouch for an instant by the door, gathering his twitching muscles for action. Suddenly he would bound across the room, leap onto the bed and chew and play with whatever parts of us were exposed. It was a shocking way to be awakened in the morning, but many times more effective than any alarm clock! There was never a question of rolling over for an extra five minutes. Even if I buried myself beneath the bedclothes for protection, Charlie would bound about above me, waiting for my slightest movement to pounce again. Eventually he would tire of the game and curl up against my head on the pillow, rumbling rhythmically a deep purr of contentment. Then I'd get up and Charlie would take over occupation of the bed.

Tess, too, was still full of the mischief of youth and she and Charlie would have some wonderful games of tag on the lawn in the evenings. They would chase each other round at top speed; then suddenly, without any apparent warning, the pursued would spin round and become the pursuer. Occasionally they collided and then an energetic wrestling bout would ensue, which would end as swiftly as it had begun, with either one, or both, leaping up and dashing away in an open invitation to be caught again.

Charlie was quickly growing from a scruffy kitten into a beautiful cat. His spotted coat was perfect. There was something about his gait, the set of his head and the regal expression in his piercing eyes that was arresting. It was like a circus on the lawn in the evenings as each of the animals, including two young hyenas, Bookie and Buster, who had recently been deposited with us, played out his youthful energy. Tim would usually avoid these outdoor games, since his nursery mates were rapidly getting too big and heavy, and in their enthusiasm they sometimes did not see him till they were almost on top of him. Charlie was also smaller than the others, but he was fast, and if the going became too rough, would leap into the safety of a tree in the corner of the garden.

As Charlie grew he began to stay longer in the bush in the evening, and by the time he was mature he had formed a regular pattern for his activities. During the day he would choose a comfortable cushion or pile of clothes in a secluded part of the house and curl up to sleep. Each evening he would make his way to the kitchen, where his meat and milk were waiting for him. Having eaten his meal, he would spend an hour or so with us all in the sitting room. Gradually one could see him becoming more restless, till finally he would walk purposefully to the dog door. Casting a last glance at us over his shoulder, he would disappear into the night. Each morning he would have returned by the time the household was up and be waiting expectantly for his bowl of milk. Then once more he would head for a quiet place in which to spend the remainder of the day.

One evening I watched as Charlie licked and cleaned a supine Tess. Then, as usual, he stood up, stretched thoroughly and began pacing about until he finally decided it was time to leave for his nocturnal wanderings. He paused by the door to glance farewell at us. "Good night, Charlie boy," I replied. "Happy hunting."

Sometime during the night there were two explosions in the bush just behind our house; undoubtedly shots fired from one of the homemade muzzle-loaders still used by the local hunters. Next

Bookie, one of the hyena cubs

morning the bowl of milk on the kitchen table remained untouched. Charlie did not come home.

Heather and I spent the day in the bush calling for him, but we neither saw nor heard anything indicating where he might be. At four o'clock that afternoon we turned for home. About a hundred yards from the house, close to the fence that surrounds Yundum Station, we heard the buzzing of flies. We found Charlie lying on his side beneath a tangle of vegetation, stiff and dead. He had been shot in the hindquarters. His beautiful face was distorted into an unfamiliar snarl and death had frozen an expression of pain and fear in his eyes. My vision blurred and I ached inside as I picked up his mutilated body. Heather and I carried him home. That evening we buried him in the corner of the garden beneath his favorite tree.

Only the night before he had playfully taken refuge in it to escape from the boisterous hyena cubs and had used it as a vantage point from which to spring back into the game. To this day it is referred to as "Charlie's tree."

In the years since Charlie died, many more serval cats have suffered similarly violent deaths. Where once his kind was plentiful there remain but a few, for their skins have become valuable to the local hunters. They are sold to the many Momadous who now walk the streets of Banjul, or wait at the airport or on the docks to sell their handbags, belts, rugs and other articles.

It is not the hunters I blame, or even the bagmen, it is the people who buy the skins, who create the demand. These skin artifacts are popular with the tourists who visit The Gambia for seven months of the year. They buy dozens of articles to take back to friends and relatives at home. Little do they realize that, in the haste for a quick and easy profit, many of the skins are only sun-dried or green-cured, and that it may take only a few months before the hair begins to fall out. The indiscriminate waste of life is heartbreaking. With so many attractive man-made fabrics and fibers on the market today, there is no need to wipe out a species for the sole purpose of obtaining its skin.

We all missed Charlie, Tim and Tess especially. When Tim came out of the wardrobe each evening he would search the house. He was so quick and lithe he seemed almost boneless, and getting beneath the cupboards and fridge was no problem for him. For the first few evenings after Charlie had gone, Tim would patter around the house, looking in every conceivable nook and cranny, then he would curl up on the back of Daddy's chair and gaze over his shoulder at whatever Daddy was doing.

The dogs and hyenas continued to romp each evening but the absence of those few extra spots, streaking in and out among them, was painfully conspicuous.

4. ABUKO

IT IS AT THIS POINT in the story that I can reintroduce William, for it was shortly after Charlie died that he arrived. I have previously described William's appearance and Tess's fascination with him. We were careful to keep the other animals away while he was still sick and timid. Once he did decide to leave his crate, taking care of him became a twenty-four-hour job, which we all shared willingly. In fact, there were times when arguments arose as to whose turn it was to have him.

Tim and William did not meet often, as during the day Tim was safely curled up in the top of my parents' wardrobe, and by the time he came out in the evening, William was usually asleep. This was just as well, for the few times that their waking hours overlapped, William proved a little too rough. I do not think William had any intention of harming Tim, but, like many human children, he wanted to pick the genet up. This would not have been so bad if he had known how to do it properly. Usually, however, Tim was retreating as William approached him, and William would try to lift him by his leg or tail. It took a few lightning nips from Tim to make William realize that he was not to touch him.

Though William must have seen Duff and the hyena cubs frequently while he was convalescing, it took him a little time to get used to personal contact with them, for they were a pretty boisterous trio. When he did eventually overcome his inhibitions, he contributed to the games just as enthusiastically as they did.

Mummy became a premature but perfect grandmother to William. She would thoroughly enjoy her stints of baby-minding. One day, shortly after William became active, she arrived home laden with brightly colored plastic bricks, floating ducks, rattles galore and other appropriate toys. William appreciated the bricks and would chew on them for hours. Known for her patience, Mummy would kneel on the carpet, chatting brightly to him and trying to interest him in building a block of colorful little houses. William took a real delight in demolishing these and would wait patiently while she painstakingly tried to balance a particularly knobbly, well-chewed brick. The longer she took to succeed, the greater seemed his joy as he leaped forward to unbalance it again. When they had tired of playing bricks, William might lead Mummy round the house on a treasure hunt. His favorite place was the bathroom: in William's opinion there was nothing so refreshing as washing yourself with toothpaste, he could even eat it if he wanted to, and toilet-paper rolls made fascinating material for interior decorations. A toilet too was an ideal paddling pool for a little chimp—there was even a special seat to cling to while he dabbled and paddled about with his toes.

His second favorite place was my parents' bedroom. He would play trampoline on the bed, roll about on the pillows and wrap himself up in the candlewick spread. As children we had never been allowed to bounce on the bedsprings or dress up in bedspreads, but

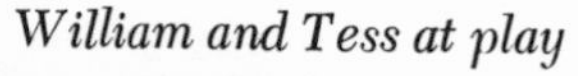

William and Tess at play

as my mother always insisted, William was different. Whenever he approached the dressing table, though, she was quick to lift him up and distract him with a trip to the kitchen for a banana, or something equally exciting. So the table remained a tempting mystery, especially reserved for some moment when everyone was busy and there was an opportunity for independent exploring. One day when my mother was in charge of William, she made the mistake of becoming too engrossed in her flower arrangement and allowed her attention to wander. It was at this point that I came in through the back door. It suddenly occurred to me that someone was missing.

"Mum," I asked, "where's William?"

She froze, her hand poised above the vase.

"Isn't he by the door, playing with his duck?" she hesitantly inquired.

No he wasn't. The green duck was there, lying desolate on its dented side, but William was nowhere in sight. As we stood gazing at the floor where William should have been, there came, from down the passage, the distinctive sound of a bottle being broken. Without hesitating, we hurried in the direction of the bedroom, colliding with William mid-passage. He looked like a hairy little Indian brave, face decorated with a stunning shade of pink lipstick, the remainder of which protruded from the corner of his mouth like a rosy cigar. On one knee he had daubed a bottle of matching nail polish, which had run down his leg in pearly rivulets that crisscrossed at his foot. His coat had been liberally dusted with a soft beige powder and the cold cream had not gone unnoticed either. To complete the effect he stank of perfume. Quite unconscious of the spectacle he presented, William climbed up my leg, put his oily arms about my neck and tightened them in an affectionate hug. What could I do but return it?

The duty of wiping up puddles was assigned to Heather and me. Every time William was seen gazing benignly ahead, legs slightly splayed, he was whisked up and placed outside on the lawn. Usually the signs were recognized too late. A puddle would form, or worse still a glistening trail would be left in the wake of whoever was hurrying William into the garden. On these occasions Heather or I would fetch the cloth from its usual place in a bucket of disinfectant on the back verandah and wipe up the puddle. Having watched this procedure several times a day for as many months, William was seen one day struggling out of the kitchen dragging a sodden cloth behind him. He approached an undetected puddle and, with both hands firmly clasping the cloth and an expression of grave determination on his face, proceeded with all his tiny might to rub the cloth backward and forward in it. His attempts at cleaning the house backfired, for by the time he exhausted himself, not only was there a trail of disinfectant from the back verandah to the sitting room but the relatively small puddle had been liberally spread to cover at least ten times its original area. He looked up at us triumphantly, and dumped the dripping cloth on the seat of a nearby chair.

As my father was normally the first up in the morning, he was usually the one to give William his cup of milky cereal porridge. William would hop about in ill-concealed impatience as it was being prepared. When it was ready he would grasp the cup with both his hands and drink as if his life depended on it. Snufflings and bubblings could be heard as he tried to drink and breathe at the same time. Milky drops would escape in haste and trickle down his chin. He was then given whatever else was on the menu for his breakfast.

When we ate a little later, William insisted on joining us. He sat next to my father. On the whole he was fairly well-mannered. He would be given a piece of bread and jam and this would normally keep him quiet long enough for us to breakfast in peace. If, however, he finished before we did, he would look around, whimpering or squeaking. Unless we were quick to respond, he would stand on his chair and reach toward whatever took his fancy, regardless of what might be in the way. More than once the milk went flying, rendering the toast inedible and reducing the sugar to a white syrup.

As William grew, it became increasingly difficult to keep him with us in the house. His curiosity and mischievousness were inexhaustible. Fond of him as we were, I had to admit that he was, and in fact still is, a very destructive child. Looking back it seems incredible how tolerant we were about accidents, Mother especially. She lost numerous irreplaceable pieces of china and crockery, and the effort involved in keeping the house organized had to be doubled, at least. William was far more destructive in the house than all the other animals put together. He was, of course, forgiven, and I think we would happily have gone on forgiving him, making sacrifices and rearranging our lives and those of the others to include him in our home, if he had not proved to us that no matter how careful we were, a household such as ours was a potentially dangerous place for an inquisitive young chimp.

I had just brought William back from an afternoon in the bush. When we returned to the house I began getting the evening feeds ready for all the animals. I was on my way to the pen to feed Bambi and Booful when Heather called out to me urgently. I reached the kitchen to discover that William had drunk some kerosene. It had been left in a lemonade bottle beside the stove by our cook, Abu. None of us knew how much the bottle had contained or how much William had drunk. I reasoned that it must taste so awful that he would have taken only a mouthful or two at the worst; however, the bottle was half full and the soaked newspaper stopper had been laid beside it, which suggested that the bottle had been full when he started. William reeked appallingly of kerosene and not long afterward crawled onto the couch. He seemed to be feeling very sick indeed.

The Veterinary Department had long since closed for the day, and as William appeared to be getting progressively worse, I rang up a doctor friend. Much to my relief he volunteered to come over immediately and have a look at William, who I was convinced was dying. While I waited for the doctor to arrive, I cradled a pathetically limp little chimp in my arms. Occasionally he struggled to open his eyes and look at me; then, while he gripped my shirt with both fists clenched, his body would contract as if in awful pain.

The doctor's bedside manner and assurances that all would be well comforted and relieved us. I was recommended to feed William as much milky tea as he could drink, but this was easier said than done. He refused even to look at his beaker and I only managed to get some of the liquid down by using a feeding bottle, on which he sucked lethargically. He spent that night on the couch with a thick towel tucked beneath him. I stayed with him and gradually he seemed to fall into a peaceful sleep.

During the early hours of the morning I dozed off too. I was awakened again at five thirty by Heather, who had come to see how William was getting on. William stirred, sat up, reached for the feeding bottle, which stood on the table beside us, and sucked on it greedily. The relief I felt was overwhelming. William, except for the kerosene fumes which still lingered on his breath, seemed surprisingly fit that morning; in fact, anyone who had not seen him the evening before would never have suspected he had felt so ill. But the incident impressed upon us that we would soon have to decide what William's future was to be.

JUST AS WE WERE BECOMING DESPERATE about finding any solution, the unexpected happened. We discovered the Abuko Water Catchment Area, two miles down the main road from Yundum Station. The catchment area was set back from the main road and completely enclosed by a fence, which was partially hidden by dense vegetation. Within the fence, at the point nearest to the road, there stood a pumping station. Nailed to the gate at the entrance was a large sign which read "NO ADMITTANCE EXCEPT ON BUSINESS." Never having had anything which could even vaguely be classed as business, we had obediently passed the area hundreds of times without stopping to look inside. We might still be doing the same if one day, early in 1968, my father hadn't been approached by Kalilu, a small farmer from nearby Lamin Village. Kalilu complained that a leopard had been killing his pigs. He offered to show us the remains of one of his animals as proof.

The same afternoon, we drove to Lamin to inspect the carcass. On reaching Kalilu's home, we were led a quarter of a mile or so to the catchment area fence. We followed this for a few hundred yards; then Kalilu dropped on all fours and wriggled through a large hole at the base of the fence.

I compare that hole, in many ways, with Alice's Looking Glass, for beyond it we discovered an incredible world we had not known existed. With each step we became more enchanted by what we saw. We had walked from the familiar savannah into the cool, damp atmosphere of a tropical rain forest. The farther in we went, the more mysterious everything became. It reminded me of the airport woods, but was even lusher. Lianas festooned the tall trees and vines formed intricate patterns against the blue sky, wherever the vegetation permitted us sight of it.

When we came to the carcass of the pig, which was old and

barely recognizable, my father seemed to find it difficult to concentrate on the matter we had come about. Finally he turned to an exasperated Kalilu and said there was nothing he could do about the leopard so long as the pigs were in the prohibited catchment area. If it began killing in the village, then there might be a case. Kalilu's complaint to us about his pigs had, in fact, been merely an excuse. Since the leopard had decided to take up residence in Abuko, he had greatly inhibited the nightly forays of the local hunters and those who went to tap the palm trees for sap for making wine or to cut down trees for firewood.

On further exploration we found the lush vegetation was supported by Lamin Stream, which supplied a perennial source of moisture. The stream culminated in a small lily-strewn lake just behind the pumping station. Having been protected as a catchment area since the early part of the century, the vegetation had been left relatively undisturbed and really gave us a glimpse of what Gambia must once have looked like. We felt strongly that Abuko should remain just as we had discovered it—a small piece of Gambian heritage, an oasis in a busy, developing country. My father sent a memorandum to the central government outlining the area's need for protection and suggesting that it be given the status of a reserve. The memorandum was sympathetically received, prompt action followed, and the Abuko Nature Reserve was established only a few weeks later, in March 1968. We were overjoyed. My father took on the additional responsibility of running the reserve, needless to say with all the assistance his family could give him.

The reserve is rectangular in shape and covers an area of 180 acres. Its lack of acreage is amply compensated for by the uniqueness and variety of the vegetation. The Lamin Stream runs roughly through the center of the reserve from north to south. The vegetation flanking the stream is lush riverain forest, which gradually gives way to the more typical savannah the farther away from the stream one gets. A narrow path was planned, to begin near the pumping station and meander up one side of the reserve and down the other in a rough semicircle, taking in all the zones of vegetation that the area held. Due to the small size of the reserve, it was impossible to allow vehicles to enter, so visitors would have to explore it on foot.

Because of the dense vegetation, the fauna of Abuko is not readily visible, though the abundant tracks show ample evidence of their existence. The easiest animals to observe are the monkeys, because they are arboreal. These include the green vervet, the red patas and my favorites, the graceful western red colobus. Occasionally, in the earlier days, we would catch a glimpse of Abuko himself, that pork-loving leopard to whom we owed so much. The year after the reserve was founded, however, he disappeared for good. I fervently hoped that he sought another peaceful pocket of The Gambia and was not shot to provide some lady with a genuine leopardskin coat.

At about the time we were discovering Abuko, Scandinavian visitors were discovering The Gambia, and tourism began to develop. All the tourist organizations then were Swedish. For us, tourism seemed a double-edged weapon. On the one hand there were obvious economic advantages for the country in developing a tourist industry—certainly it would help to support Abuko. On the other, tourism contributed to the destruction of the wildlife. Apart from the handbag trade, there were, at first, tourist hunters. Some of the tourist agencies advertised big game hunting as an attraction in the package holidays they offered. Enthusiasts arrived in The Gambia with powerful rifles, only to find that we did not possess any big game. On those first hunting trips, anything that moved was shot. Instances were reported of hunters shooting our colorful birds —apparently just for the sake of firing a rifle, for the birds were left where they fell. Happily, since my father brought this and other incidents to the notice of the Gambian government, hunting by tourists is now prohibited.

It was not precaution enough to have Abuko designated as a protected reserve on paper alone. For the first couple of years we could not afford to employ guards or nightwatchmen. We thought the best policy was to encourage the people of Lamin and other villages close by to understand what we were trying to do. Our biggest hazard was, of course, the local hunters. One of them, named Doodoo, we knew quite well. Doodoo hunted for meat, which he sold in the village. We had no complaint about this. Hunting for food was one thing; it was when the hunters began killing for skins that we became worried. My father called at Doodoo's house one evening and asked if we could arrange a meeting of all the hunters in the immediate vicinity. During the meeting my father explained exactly why he wanted Abuko protected, and asked for the hunters' help. Surprisingly, they all agreed with him, and many of them swore on the Koran that they would not attempt to hunt in Abuko. Some of them went even further and volunteered to act as guards, offering to inform us if they heard of anyone poaching there. They have kept their promise, and to date no one has violated the undertaking.

Later, when the Abuko Orphanage was founded, it became necessary for my father or myself to go up each evening to feed the animals. Each time we would take with us a few of the children who were waiting at the reserve gate. To ensure that we took only those who were genuinely interested, we asked them to bring a small amount of fruit or grain as their admittance fee. Eventually the children sorted themselves out into groups, one group for each day of the week. At first they were more of a hindrance than a help and some of them tried to tease the animals, but our attitude toward the orphans was gradually absorbed. Their personal contact with the animals was, I think, invaluable. The children were learning not just to see food when they saw an antelope but to appreciate the animal's beauty and its right to existence. I would allow them to help me bottle-feed, and they seemed to derive great pleasure from doing this. Since Abuko's future will eventually lie in their generation's hands, I felt it vital that they should understand why we wanted to preserve the land and its inhabitants.

To publicize Abuko and to try to combat the sale of animal skins, my father began giving slide shows in the hotels. The money raised at these lectures was spent on the reserve. If we happened to have any suitable orphans in our care at the time, we would take them with us, to win over our audience with the charms of the real thing. At one time we were able to take along two little serval kittens. Two more heart-winning creatures one could not wish to meet. To emphasize this, I would bring with me two feeding bottles and a glove, and a member of the audience would be allowed to come out and feed them under our guidance. The kittens did not care much for being confined in a basket, but they were very effective ambassadors for their species. I know that they made several

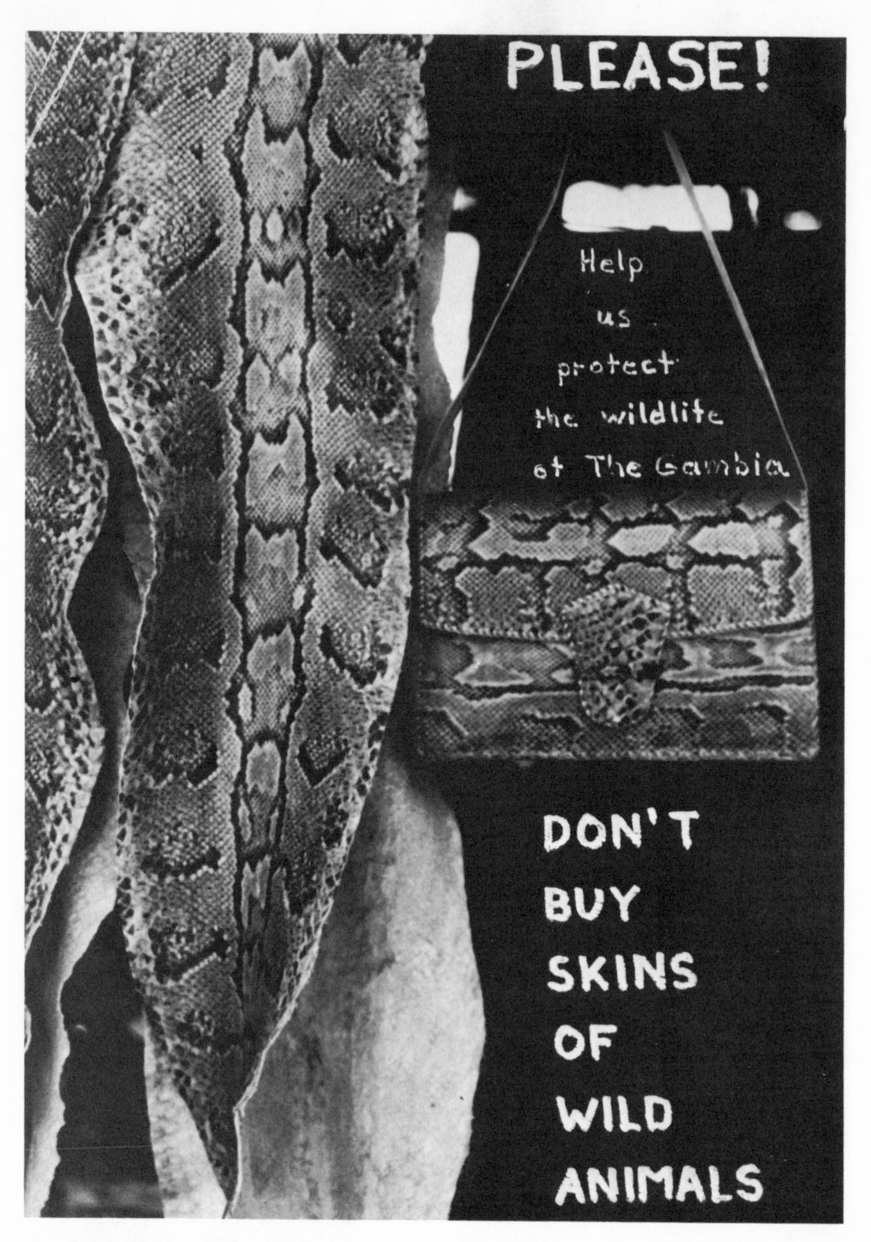

A poster from the early days of Abuko

lasting converts, and they continue to do so by gracing Abuko with their presence to this day.

The easiest animals to exhibit were those whose charms and fine qualities were *not* so obvious to our audience, such as the royal pythons that we managed to rescue before they became wallets or handbags. The royal python is the "gentleman" of the animal world, almost incapable of any kind of aggressive behavior. If alarmed or uncertain, he politely curls himself up into a ball with the most vulnerable part of his body, the head, safely tucked away in the center of his coils. He quickly responds to gentle handling and it takes only a short while before he feels confident enough to remain extended when he is touched or held.

Percy, a beautiful specimen about five feet long, became quite a household pet and, until his release into the reserve, made regular and impressive appearances at my father's talks. It was amazing how many people recoiled in horror when Percy was initially produced, but after watching him wrap himself around our arms and shoulders and explore the unfamiliar surroundings with a flickering, inquisitive tongue, they would gradually relax. I always felt Percy had scored a big victory when some of these people eventually came forward to stroke his fine skin. They would look at him in amazement and exclaim that they had always thought snakes to be slimy

Stella and Percy

and cold. On the contrary—the skin of a healthy python has one of the finest textures with which I have ever come in contact.

One evening, during one of these talks, I was sitting unobtrusively near the exit, with Percy confined in a zipper holdall beside me. For some reason that evening he was particularly active and, halfway through the talk, began to emerge from the corner of the bag, having somehow or other pushed the zipper backward to make an opening. Daddy was talking enthusiastically and his audience was silent and attentive. I knew from experience that an uproar could be caused if Percy were to make his appearance unannounced, so I picked up the bag and slipped out of the room.

The ladies' powder room seemed the obvious place in which to allow Percy the chance to burn up some of his excess energy. There was an elegant elderly lady combing her hair in the mirror, so I hastily slipped into one of the cubicles and closed the door. Presently I heard the staccato clip-clop of feminine shoes and the door of the ladies' room opened and closed as the only other occupant, I thought, left. I bent down to release Percy. Unzipping the bag, I found he seemed intent on reaching the floor. Not wishing to alarm him, I did not tighten my grip on his writhing body and he succeeded in escaping. He promptly disappeared under the six-inch gap between the cubicle door and the floor, leaving me alone. By the time I had unlocked the door he was already well down the room and was about to enter the last cubicle in the same fashion as he had escaped from the first. Fortunately he hesitated for a second, which gave me time to catch up with his tail.

Just then I heard the sound of someone approaching. The most sensible thing seemed to be to release that part of Percy still in my possession and follow him into hiding, and this I proceeded to do. I pushed on the door; to my horror, I found it locked. I was still pushing incredulously when the elderly lady re-entered. She must have wondered why I was trying to open the door to the only locked cubicle in a row of five empty ones but smiled at me anyway. I managed a weak smile in response. Suddenly a scream erupted from the locked cubicle. At the top of my voice I began trying to console the invisible victim. I begged her to believe that the snake was harmless, his name was Percy and she was in no danger. As suddenly as it had begun, the noise abated.

"Ma'am?" I hesitantly inquired. "Are you all right? Would you just open the door for me, please?"

I tried again to explain. I had not gotten very far when there was another scream. This time it was my turn to jump, for it came from the elderly woman behind me; she was flattened against the wash basin, staring at my feet. I looked down and to my relief saw Percy re-emerging. I bent down and picked him up. Perhaps in reaction to the screaming, he was once again his old tractable self.

"You see," I began, "he's quite harmless and very gentle. . . ."

The lady facing me remained unconvinced. Suddenly there was the sound of a bolt being drawn back and the door beside me flew open. A young fair-haired girl scurried out, looking dazed, floral panties knee-high. She took one look at me and Percy, yanked the panties up to their proper place and fled.

There was silence for a moment. When I glanced at the woman facing me, she was chuckling. We arrived back in time to hear the end of my father's talk.

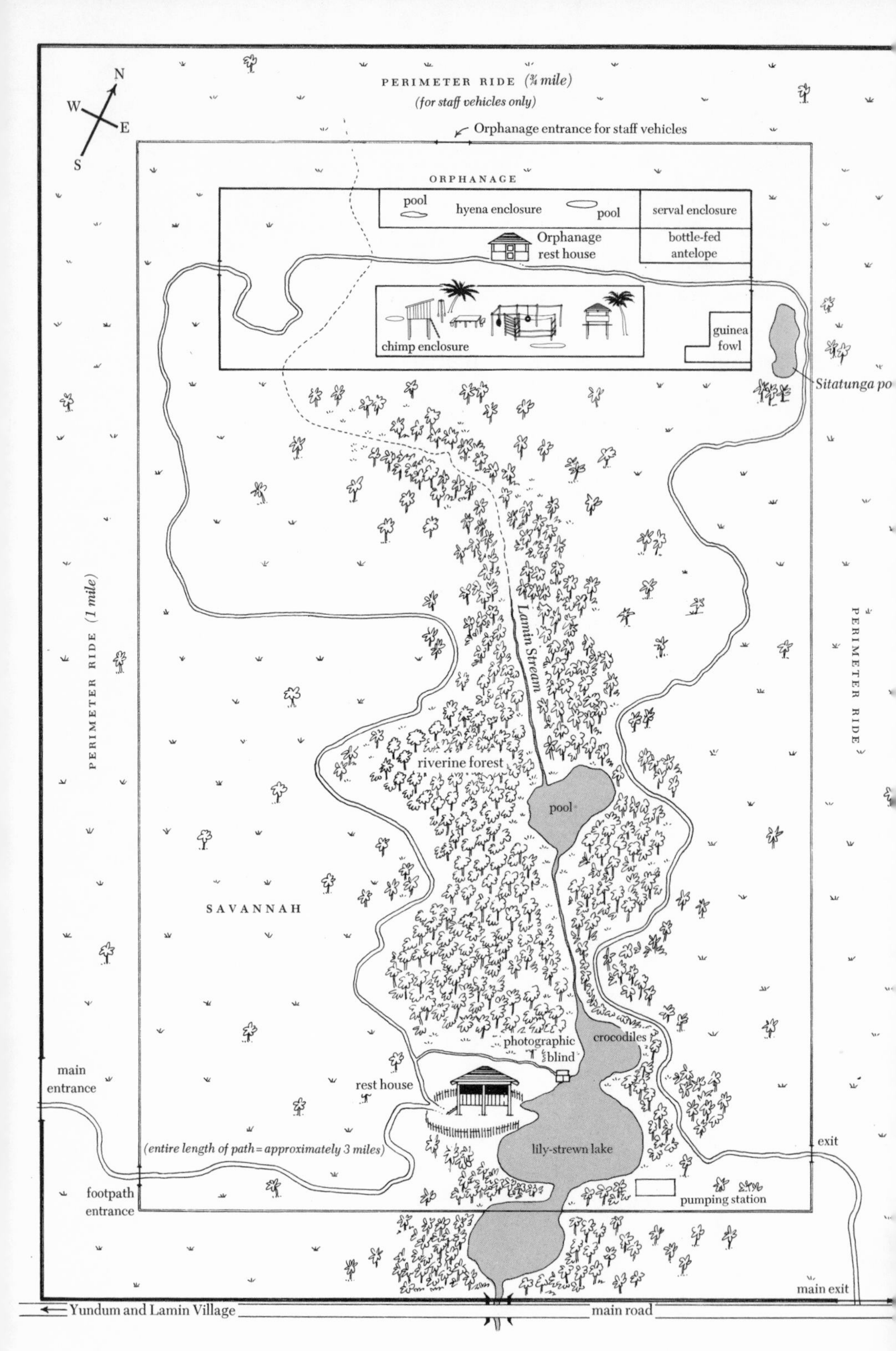

THE ABUKO NATURE RESERVE

5. THE CHIMPS ARRIVE

IT WAS A FEW MONTHS after the discovery of Abuko that Ann arrived. Ann was a female chimp whom we estimated to be about one year younger than William. These two turned out as different from one another as any two chimps could be. Ann was quiet and unassuming and yet, when necessary, she showed a very determined streak. For one who was apparently so fragile, she stood up admirably to William's robust play and occasional bullying. She was indifferent to the hyenas and to Duffy and they soon appreciated that she didn't really want to join in their games. Tess's attentions she happily tolerated. There was something in her manner which I find difficult to describe. She was distant, composed and self-sufficient. She commanded a respect from William and, later, from her other chimp playfellows which they rarely showed to anything or anyone else. Beneath her composure she had a lively, intelligent mind. Unlike William, who boasted about his achievements and was openly mischievous, Ann was far more discreet and carried out her activities with something approaching cunning.

Chimps do not occur naturally in The Gambia and we could not understand at first why they had suddenly begun to appear. William had been the first chimp we had ever seen outside a zoo. Later we learned that just before William found us, a male chimp had been bought for an exorbitant price in Basse, a village two hundred miles up the river, close to the Gambian-Senegalese border. The news had obviously reached Guinea, where we believe our chimps were captured, and the rush to supply this lucrative new market began.

Ann was the answer to our prayers. With the discovery of Abuko we realized that we had a perfect home for William; however, to put him there by himself, after he had become used to being a member of a crowded household, would have been unfair. With Ann as a companion there seemed a chance that they might make the change. Until Abuko was ready, we built temporary accommodation for the two of them in the garden. This was a large, airy cage of about 750 square feet, which we filled with tires, ropes, complicated climbing apparatus and a box full of William's toys. I think my father got as much pleasure out of these amenities as the chimps did, for every time he went into the garden he would be missing for a while. Usually I discovered him swinging about in the cage, his excuse being that he was testing the apparatus.

As long as someone remained with Ann and William in the cage they would amuse themselves for hours, inventing various ways of using the material at hand. However, if we left them alone they were desperately unhappy. We tried leaving Tess with them once or twice, with the result that William's screams of protest were punctuated by mournful canine whimpers.

Ann was my baby from the first and, unlike William, found it difficult to accept the whole family as foster parents. It took me a long time to win her confidence, and when she did accept me, I felt almost honored. As I had done with Kim, the red patas monkey,

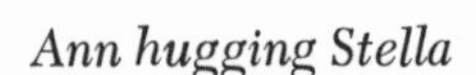

Ann hugging Stella

years before, I would strap Ann to my back with a length of cloth, which at least left my arms free for William. She seemed quite happy with this arrangement.

Gradually I accustomed them to spending more time by themselves. William's crate, which contained his bedding, was moved from the verandah and raised on poles in the cage. Their supper was served near the newly relocated sleeping quarters. At first I remained with them until they slept, both of them usually staying steadfastly on my lap as dusk approached. With stomachs round and full, they would fall asleep in my arms. Carefully I would try to place them in the crate without waking them. As a rule I was only partially successful. They would both half wake, but on seeing my face and hearing my voice, they would clutch the nearest warm thing to hand, which happened to be each other, and promptly fall asleep again. Eventually it became possible to leave them both as soon as supper was over. They protested at first but soon accepted the idea, and so long as I was out of sight they would amuse themselves happily before climbing into their sleeping box and curling up together.

We had decided that in the center of the reserve we would fence off an area of twenty acres which we would call the Orphanage. Once the animals in the house were weaned, they could be accommodated here, where they would be able to get used to a more natural environment while still under close supervision. In a sense, it was a mini-rehabilitation center.

We spent several evenings making plans for a large outdoor enclosure to house the chimps for part of the day. The idea was to enclose about half an acre of bush with eight-foot, plastic-covered chain-link fencing. On top of the wire there would be an additional three sheets of smooth corrugated iron set horizontally one above the other. In this way the chimps would be able to climb to the top of the chain-link fencing but no higher.

William and Ann had watched their future home grow day by day. Ann had remained on my back most of the time but William, ever anxious to be in the midst of whatever was going on, had taken an active, if ineffectual, part in the work. By the time the enclosure

William at work

was finished he had mastered the use of the hammer and pliers and was well on his way to conquering the spade. Though he understood the basic principles of its use, the spade was almost twice as tall as he was and too large and cumbersome for him to manipulate easily. I bought him a child's beach spade, which was more his size, but he insisted on using the tools of his workmates.

The chimp enclosure took months to build, but as it neared completion we all felt that it had been a job well done. Large trees close to the fence had to be removed to prevent escape, but otherwise the enclosure was left with as much natural plant life as possible. A complex of jungle gyms was constructed which included

William on the climbing apparatus

climbing frames, suspended tractor tires, a web of high-tension wires and a primitive seesaw. At either end of the enclosure stood a square wooden hut on stilts with overhanging slatted roof. One of these had been left with no walls and would serve as a shady resting place during the heat of the day; the other was enclosed on three sides. A wooden ladder led to the fourth, open side. Inside we had hung two hessian hammocks for the chimps to sleep in. Between the two huts we dug two circular pools to serve as drinking places. Beside one of the pools a solid wooden table and two benches were cemented into the ground, where we intended feeding the chimps.

We were just putting the finishing touches on the enclosure when we heard that two more chimps had been brought into Banjul. They were a male and a female and had been bought by a European store manager. Since they were approximately four years old and freshly caught, they were unmanageable and had therefore been enclosed in a cage behind their owner's shop. Here they were easily accessible to all those who wished to approach and tease them through the bars. Being in a town, they were particularly vulnerable to the merciless provocation of street urchins. These children would interrupt their begging in the streets for an amusing hour or two in front of the chimps' cage. They found it hilarious to make faces at the chimps, to spit at them, to poke sticks through the bars and to proffer food which they would snatch away again the moment the chimps reached out to accept it.

We also discovered three more young chimps for sale at the back of the Albert market in Banjul. Two of them were very young and in terrible physical condition. They were suffering from malnutrition and their coats were filthy and matted. One of them had an infected sore on his head and both had running noses. The third was older, perhaps four years old. Though not emaciated, he was dejected and miserable—sitting hunched and still at the back of his soiled, rusty mesh cage. When I asked the trader where the chimps had come from, he said that they were from Guinea and that he could get me as many as I wanted. "But," he went on, "this kind of animal very dear, cost plenty money."

I knew that if the two little ones were not rescued and looked after properly they would die in the next few days. The temptation to beg Daddy to buy them was unbearable. My father was just as concerned as I was, but even if we could have afforded them, purchase was not the answer. It would only serve to encourage the trade, so that each time we bought a chimp we would be indirectly murdering another five or six who would be caught in order that just one more would arrive in The Gambia alive to replace it. How many of them had already died on the journey, how many mothers had been shot, so that baby chimps could be brought to The Gambia and sold?

We knew that in Senegal the chimp was a protected animal and that it was illegal to keep one without a license from the Département d'Eaux et Forêts. We suspected that an attempt was being made to establish a chimp trade in The Gambia, where the regulations for the protection of wildlife were laxer. Although there were laws concerning cruelty to animals and a list of protected animal species, The Gambia did not at that time have a Conservation Department, so the frequent violations of these laws went unnoticed.

My father decided to appeal to the Inspector General of Police. We were relieved at his quick and positive reaction: the chimpanzees in The Gambia were to be confiscated and sent to Abuko Nature Reserve. We rushed back to the market, but only the juvenile male was still there. Both the little ones had died the previous day.

We called the surviving chimp Albert. He had not been in captivity long and was still very wild. He was about 28 inches tall and weighed 35 pounds. A fresh scar curved over his upper lip, which I suspected to be the result of human maltreatment. The cringing, desperate fear he displayed when anyone tried to get close to him seemed to confirm this. We hastily prepared a guinea-fowl hutch in the Orphanage, and it was here that Albert lived for the next few days till the enclosure was complete.

I spent hours trying to alleviate the loneliness and desolation I thought Albert must be feeling by taking William and Ann over to his cage. Realizing that I frightened him, I would sit quietly and read in the hope that he would begin to interact with William and Ann through the wire. When this seemed to be having no effect, I began taking food to the hutch to offer him. Even when he was able to see the other two gorging themselves, he showed not the slightest inclination to come to the wire and share with them. He would sometimes watch me morosely, but there seemed nothing

William, Ann or I could do to entice him out of his apathy. He remained a distrustful and pathetically lonely figure.

As soon as the enclosure was completed we tried to move Albert in. It was a problem to know how best to accomplish this—he was so wary and frightened that no one could approach him. Finally a large crate with a guillotine door was prepared. We filled it with fruit and placed it at the entrance to the hutch. The sliding door of the crate was opened and held in position by a pin. A long string was attached to the pin so that we could close the crate from a distance.

Albert was terrified of the crate. I hid with the string in my hand for many hours before he finally approached it. He peered in at the abundance of fruit and I sat tense, waiting for him to enter. Then in one lightning movement he ran in and snatched a fruit. As he leaped clear again, his back knocked against the delicately positioned pin and the crate door slammed down behind him. He ran to the end of the hutch, screaming, then turned to see that the gap between the top of the hutch door and the crate, covered by the crate door when it was open, was now exposed. He hesitated only one brief second before leaping through the gap and racing for the dense cover of the main reserve.

We searched for him but Albert had vanished. Catching him would have been almost impossible anyway. It would have taken Albert only a few minutes to reach the perimeter fence of the reserve. Once over that, he would have been exposed to a thousand dangers, the biggest of which would be the farmers protecting their crops. I wondered miserably what would become of him.

Soon after Albert's escape, the European store manager who kept the two chimps behind his shop telephoned us. The female "monkey" had died. He thought the male too dangerous to be interesting and asked us if we would take him.

When we reached the shop I went round to the back to see the chimp. He looked a solid individual, weighing about 35 pounds and was 28 inches long. Unlike Albert, he carried no visible scars or signs of serious physical maltreatment. His coat was dry but long and abundant. He had protruding ears which seemed unusually large and rather flaccid. It was the paleness of his otherwise tough-looking face that struck me most. His name was Cheetah. I guessed him to be about 4 years old.

Cheetah was driven straight to the reserve, and this time we successfully managed to get our chimp into the enclosure. When the door to his cage was opened, Cheetah hesitated for an instant, unsure of what was happening, then bolted out into comparative freedom. He took a brief drink from one of the pools and watched from a distance as his old cage was taken away.

That evening I took Ann and William to the reserve with me, impatient to see Cheetah's reaction. Cheetah started grinning the moment he spotted them and extended his hand trustingly through the wire. The introduction looked so promising that I decided to take Ann and William into the enclosure. Cheetah seemed nervous when I entered and, though showing all the signs of desperate curiosity, kept well away until I sat on the feeding-table bench and William climbed off me and onto the ground. Ann remained on my back, clinging tightly to my shirt.

Despite being somewhat larger, Cheetah approached cautiously —alternately panting his submission and drawing his lips back to squeak uncertainly. William, I knew, was equally nervous but, reassured by the fact that I was sitting right behind him, he fluffed out his sparse coat to look as large as possible and bravely advanced toward Cheetah. For an instant they stood inches apart, both with nervous open grins, screeching in uncertain excitement. I knew that one aggressive sound or gesture from either of them could wreck the new relationship and cause a fight, but fortunately William half turned round and presented his white-tufted bottom in a sign of submission. Cheetah immediately responded by embracing William.

At this point Ann climbed off my back and stamped her feet on the bench in a state of anxiety for William mixed with general excitement. Her coat bristled, but as Cheetah began to groom William, she plonked herself on my lap and, firmly gripping my hand, regained her usual wide-eyed expression and became a passive audience once more. William and Cheetah now began to play. Cheetah was occasionally a little rough, but William was clearly enjoying every minute. Cheetah ignored my presence as much as he could and, to my relief, showed no tendency to act aggressively.

Throughout the ensuing week I took Ann and William into the

Cheetah

enclosure whenever the opportunity arose. I was patient and let Cheetah make the first moves in establishing our relationship. He was soon accepting food from me and allowing me to touch and groom him, but he was still a long way from completely trusting me.

Once he felt confident about Ann and William, Cheetah began approaching me more often. Each time he would take one or two extra liberties, testing me to find out how far he could go toward intimidating or dominating me. At first his intentions were disguised in play. He would pull my hand, pinch my leg and pretend to bite me. Gradually the play became more serious. I had watched the same tactics being used by William and Ann when they met new people, so I was not entirely unfamiliar with them. The time came, as I expected, when I had to stand my ground and assert some authority. Cheetah had been in the reserve just over a week when, one afternoon, I gave the chimps some bananas. William and Ann were given four each, but as Cheetah was bigger he got six.

Cheetah wolfed down his fruit and had finished it all by the time Ann was delicately beginning her third banana. Cheetah strode boldly toward her. She guessed what he had in mind and grasped her remaining banana firmly in one foot. Thwarted, Cheetah began to whimper. Soon the whimper rose to a harsh scream, and Ann hurriedly scampered onto my knees. Still nibbling her banana, she calmly observed Cheetah's show of temper from her refuge. With scarcely any warning Cheetah leaped at my leg and bit it hard. Instinctively I kicked out. Cheetah fell backward but he sprang to his feet and leaped at me again, this time ripping my shirt and sinking his teeth into my arm. Almost without thinking, I found I had leaned forward and bitten Cheetah equally hard on the shoulder. William grabbed Cheetah's foot and bit it too. Ann screamed and snatched at Cheetah's face. He released my arm and ran screaming to the end of the enclosure. William and Ann climbed onto my lap and hugged themselves to me.

I allowed Cheetah to scream for a moment and then approached him. He looked up at me, confused and frightened at my sudden violence; since his arrival he had received only kindness and warmth from me, something he may possibly not have experienced since losing his mother. At that moment, when all the world seemed suddenly to have turned against him, I knew that more than anything he wanted comfort and reassurance.

I squatted down in front of Cheetah and held out my arms. I could almost see the struggle in his face. Would I attack him again or give him the physical comfort he so badly needed? The more undecided he became, the greater grew his desire for reassurance. I reached out and touched his head, and he responded by throwing himself against me and hugging me tightly. I returned his embrace and spoke soothingly to him. For the first time I lifted Cheetah. He sat on my lap and panted good-naturedly as I carried him to the table and gave them all some more bananas.

That afternoon the seeds were sown for the confident and deeply affectionate relationship we later shared.

Three days later I took William and Ann to the reserve for their usual afternoon play session with Cheetah and left them there for the night. There were dreadful protests when I tried to leave them in those strange sleeping quarters and it was well after dark before I was able to creep away. I am still not sure who was the most upset that evening—William, Ann or I. Early the next morning I crept into the Orphanage as quietly as I had crept out of it the previous evening. To my delight all three chimps were engrossed in play. Even serious little Ann was rolling around in the brawlingest rough-and-tumble I'd seen for a long time. She was covered with bits of grass and dead leaves but was obviously thoroughly enjoying herself.

Despite all the amenities in the enclosure, we still felt that it would be boring for the chimps to have to spend all their time in it. A routine was established whereby each morning I would take all three of the chimps out into the main reserve for at least three or four hours. I would carry William and Ann out of the enclosure while Cheetah followed. They were free to climb, to play, to feed or to rest. I never had the slightest worry that I would lose any of them, for the moment I moved or walked away from where they were

William and Bambi

playing, all three would rush to my side and climb all over me or clutch at my hands. Each morning I would lead the chimps to a different site. At first I used to take a small picnic basket of fruit and drinks, but I quickly abandoned this and concentrated on trying to teach them to climb and to make use of the abundance of edible foodstuffs that the reserve contained.

We were frequently accompanied on our walks by Bambi, the harnessed antelope. Bambi would occasionally stop to nibble at the infinite variety of leaves, flowers and grasses, then canter along the path to catch up with us. She actually played with the chimps, splaying her long delicate legs and lowering her head into their chests, then skittishly sidestepping and backing around them in unmistakable high spirits. Ann and William especially used to pull her ears or topple and slide on her back and legs. Any other antelope I've met would have been thrown into panic by this treatment, but Bambi didn't bat her beautifully lashed eyelids. So relaxed was she with the chimps that she would lie down near us to rest and allow Ann to lean against her side or shoulder, with small woolly arms clasped round her neck.

One morning as I was walking back toward the rest house after one of these excursions, I heard my father calling me. I turned to look in the direction of the guinea-fowl hutch where Albert had been housed and saw that a group of the Abuko staff and my father were gathered about it. With quickening anticipation I wondered whether Albert had been found, and hastily made my way over there.

The floor of the hutch had been thickly bedded with clean straw, in the midst of which lay a very sick-looking young chimp. He was lying on his side, so I did not immediately see the cause of his suffering. From what I could glimpse of his face, it appeared sunken and sallow, almost yellow, and the one eye visible had a definite upward slant to it. I couldn't help thinking that he looked remarkably oriental, and I decided to name him Wong.

My father told me the police had found him in Banjul and had confiscated him. Apart from anything else, he had a terrible ulcer on the side of his face. It was unlike my father to be pessimistic, so I was startled when he added quietly that he felt there was little hope for the chimp's survival. At that moment the black form on the straw stirred and painfully arranged his bony limbs to attain a sitting position. For the first time I saw the left side of his face and realized despairingly what my father had meant. Wong's entire cheek had been eaten away by ulcers, leaving gaping holes which revealed his gums and teeth. The vet arrived shortly afterward. He looked horrified when he saw the chimp and stated honestly that he had little hope of saving him. However, we agreed to try for a couple of days to see if Wong would respond to treatment.

For the next two days I spent almost all my time in the hutch with Wong, trying to revive in him some interest in life. I groomed him for long periods in the same way as I imagined his mother would have done and managed to persuade him to try some banana and other soft foods, but eating for Wong was a pain-filled process, and much of the food slid out of his mouth through the openings in his cheek.

Poor Wong, it was little wonder that he looked so helpless and lonely. Just when I felt he was beginning to trust me, it would be time to inject him again or to clean his tormented face, a process that was bound to hurt him. I was sure that, every time, he felt his trust had been betrayed. Most of the day he would ignore my presence, lying still and dejected on the bed of straw. If he wasn't sleeping, he would gaze through swollen eyelids at the forest beyond the cage. The expression in his chestnut eyes was extremely sad. Sometime during the lonely hours of his second night at Abuko, Wong died. He lay with his thin arms stretched before him as if his last instinctive need had been to embrace someone. I wished so much that one of us had been with him to respond. Though he was distrustful of humans, perhaps in those hazy last moments of life he might have found the comfort he had yearned for throughout his suffering. Perhaps he would have believed he was holding his mother once again. But we hadn't been able to do even this small thing for Wong.

6. TINA, THE FOSTER MOTHER

As THE RESERVE BECAME MORE POPULAR and received an increasing number of visitors, we found it was taking more rather than less out of the already strained family coffers. Since it was officially a government project, all the entrance fees were government revenue; but even if the government had been able to allocate all these funds to the development of the reserve, it would still have been skimpy fare in those early days.

The fortnightly allowance for feeding the animals in the Orphanage was rarely enough. There were periods when at least twice as much had to be found from another source to meet the food bills. During the tourist season the donations received from the lectures helped, and many of our friends would send their excess garden produce to us for the animals, but even so most of the additional funds needed for the reserve came out of my father's very ordinary government salary.

During the summer rains the reserve was closed to visitors and the paths were often waterlogged. We took advantage of this respite to continue development work. With the incessant rain there was a sudden acceleration in the rate of growth and the reserve became more than ever a cool, green tropical jungle.

I had waited for a shower to stop before taking out the chimps. They were full of the joys of life in the refreshing dampness after the rain. Ann clung to my back, William and Cheetah cantered and somersaulted their way down the path ahead of me, sometimes wrestling together or chasing one another up and down the hanging vines. A group of colobus fled from the trees closest to the path and sat half-hidden among the leaves a safe distance from us. Long, red, ropelike tails hanging straight among the vines were all I could see of many of them.

Cheetah and William paused to look up at the monkeys, then chased each other round a bend in the path and disappeared from sight. Almost immediately I heard Cheetah give hoots and grunts of recognition. Abduli, our warden, came in sight with a chimp pulling at each of his hands—he was slightly out of breath and had obviously been hurrying. "Stella," he burst out, "I've just seen Albert."

"Albert," I echoed. "Albert! Are you certain?"

He assured me that he was. Abduli carried Cheetah for me so that we could hurry down the path to where the chimp had been spotted. When he stopped I followed his gaze. High in a large mampatto tree just ahead of us sat a young chimpanzee. There was no doubt that it was Albert. He looked healthy and well-nourished and, in fact, sported quite a paunch. Cheetah, William and Ann were openly curious, but I could detect no excitement or inquisitiveness in Albert's manner—he merely sat and serenely surveyed the group below him. I sat down so that Ann and William would feel free to leave me. Cheetah led the way, but as he drew close, Albert made an expert aerial crossing into a neighboring tree and vanished. Cheetah looked after him, then decided to come back to me. Ann and William followed him.

Six months had passed since Albert's escape and there was no way of telling how many times he must quietly have watched us. It was amazing that he should have remained in the reserve at all; even more incredible that he had until that day kept his presence so secret.

From then on we all kept a lookout for him, but it was another two months before he chose to reveal himself again. Each time he was seen, he was silently watching the men at work clearing the paths. During his first year in the reserve he remained totally independent—not once did he make any attempt to approach or associate with William, Cheetah and Ann.

One day, 8 months after we'd installed all the chimps at Abuko, I was told that another chimpanzee was in town. I rushed to the spot and was ushered into a gloomy corrugated iron shack, in one corner of which was a wooden box. I stepped over to it, and immediately a strong smell of chimp dung stung my nostrils. Peering through a crack in the crate, I saw a weeping brown eye staring back at me, and through other cracks there poked tufts of black hair.

With the help of the police, it took little time to arrange for the crate to be confiscated and removed to Abuko. We discovered the chimp was a female, the biggest that we had so far acquired. She weighed about 40 pounds and was about 3 feet tall. She had already lost her incisor milk teeth, and the large white permanent ones were in the process of erupting to replace them. We guessed she was about six years old. One of the front top teeth was markedly larger than the other, which gave her an odd, gappy-looking grin. Apart from being thin and having a slight infection in one eye, there seemed little physically wrong with her, so we released her directly into the enclosure with the others.

Surprisingly, when the slatted lid was removed from her box, Tina, as we called her, did not leap for freedom. She drew her legs back uneasily when the other chimps approached and screamed her fear at the confusing new situation. It was several minutes before she attempted to leave the crate. None of us knew how long Tina had been confined in the crate—where the only position possible was a cramped sitting one—but it cannot have been less than a week, for when she tried to leave it, she had serious difficulty walking. In a painfully stiff manner she made her way to a small tree and sat down in the shade beneath it.

An hour later she managed to climb into the crown of an oil palm in the middle of the enclosure, and there she remained for the next three days. She watched when I fed the chimps that evening but showed no inclination to come down and retrieve any food for herself. I tried throwing some oranges up to her, and one or two of them caught in the fronds. These she ate immediately, so I threw her some more. Half an hour later I had an aching arm but Tina, who was now actually leaning out and catching some of the food, had eaten a fairly substantial supper. It was thus that we kept her supplied with food for three days. During this time she never attempted to leave the palm tree. She had made an enormous nest in

Tina in her crate

the crown of it and I think she must have slept most of the day, for she was rarely visible.

On the fourth evening she carefully descended and attempted to share the food on the table with the other chimps. She was timid in her approach, and though she was bigger than any of them, Cheetah had only to bristle his coat and stamp his feet on the table to make her withdraw screaming. To my disappointment neither William nor Ann was friendly toward the newcomer. Finally, after much hesitating, she managed to grab a loaf of bread and stiffly climb back to her nest at the top of the oil palm.

That evening over our own supper we held a family conference to discuss what the best solution for Tina would be. We came to the decision that we should risk giving her the freedom of the reserve. She was unhappy in the enclosure and, still being so timid, would cause no problems with tourists if we gave her her liberty. Also, she was potentially an ideal companion for Albert, whom we still saw occasionally. The question was, would she remain in the reserve?

On the fifth morning after her arrival I left the enclosure door open when I took out Cheetah, William and Ann. When we returned Tina had gone. She did not keep us in suspense long as to whether or not she would stay. A couple of evenings later she was seen sitting in a tree close to the back of the enclosure. I piled a tray high with tempting fruit and some bread and placed it on the ground as close as I dared to where she sat. I then retreated to the rest house to watch how she would respond.

Bambi, who also happened to be present, considered this too good an opportunity to miss. She made a beeline for the tray and began to help herself. Tina watched. Slowly, and with extreme caution, she began to lower herself to the ground. In a series of sudden dashes, punctuated by pauses during which she checked to make sure that there was no hidden trap, she approached the tray. Ignoring Bambi, she proceeded to load up with as much fruit as she could carry. She placed an orange on her neck and held it there by leaning her cheek against it, pressed some more fruit between her thighs and lower abdomen, put a loaf in her mouth, filled her arms to capacity, and gripped a banana in each foot. Thus laden she hobbled back toward the safety of the dense vegetation. She was forced to walk on her heels in a crouched and huddled position. Pieces of precariously balanced fruit kept dropping, and as fast as she recovered one, another piece fell. She finally reached her chosen refuge with amazingly low losses and settled down to eat.

A precedent had been set. From then on she appeared faithfully in the Orphanage each sunset to receive her quota of fruit. Bambi fell into the habit of sharing the meal with her, and even after a special elevated feeding table had been built among the vines and the antelope could no longer reach the food, the two of them were

Tina and Bambi

Pooh, about one year old

frequently seen together in the reserve. Bambi would stand beneath the trees where Tina fed and eat all the fruit she dropped. Tina would also groom Bambi and keep her ears and face free from the bloodsucking ticks which attach themselves to antelopes. For Tina in those first lonely months, Bambi was a valuable companion.

One Sunday morning I was pottering around when a young French couple walked into the Orphanage. In the arms of the woman sat the smallest chimp I'd ever imagined. He was still so young—about 6 months—that he couldn't walk, and the movements of his head were uncoordinated and jerky. From his chest down he was enveloped in an extravagant pair of frilly pants from which poked his legs and miniature clenched pink feet. The hair on his head stuck up and out like a golliwog's and set in his tiny face was a pair of brown eyes which wore a permanently startled expression. He was the most charming little creature I'd ever seen.

The couple had found him in Basse while on trek upriver. He'd been lying beneath a mango tree in a native compound and apparently was being fed only water. They had bought him and he now lived with them in their flat in Banjul. I later learned that Pooh, as the baby was called, lived in the lap of luxury—he had his own mosquito-netted cradle and a special high chair and was the apple of his foster mother's eye. The couple brought him to the reserve occasionally on weekends, but he never showed much interest in the other chimps.

Although we would have liked to add Pooh to our team, the French couple were not yet ready to part with him. Another recruit soon arrived, however. In the several months since Tina's confiscation, there had been no further signs of new chimps in Banjul. We thought that perhaps the traders had learned that The Gambia wasn't such a good place for trading after all. Then one evening we found someone waiting outside the reserve gates with a large covered basket on the ground beside him.

The man could speak no English or any other language that we knew, so communication was difficult, but he conveyed his message by opening his basket carefully and revealing its contents. At the bottom of it sat an infant chimp not much bigger than Pooh. The basket was clean and the chimp looked in good condition. We motioned to the man to put his basket in the Land Rover and accompany us to the Orphanage, where we hoped Abduli might be able to interpret for us. Luckily he and Abduli did have a language in common, and we explained to him the rules about bringing chimps into The Gambia. We mentioned that three chimps had already been confiscated by the police and sent to Abuko. We even tried to make him see why confiscation was necessary and how cruel it was to take the little chimps from their mothers. Then we gave him the money for his journey home and he left us, moderately content.

We called the new chimp Happy. He was an exceptionally

Foster mother Tina teaches Happy to eat a rhun-palm fruit.

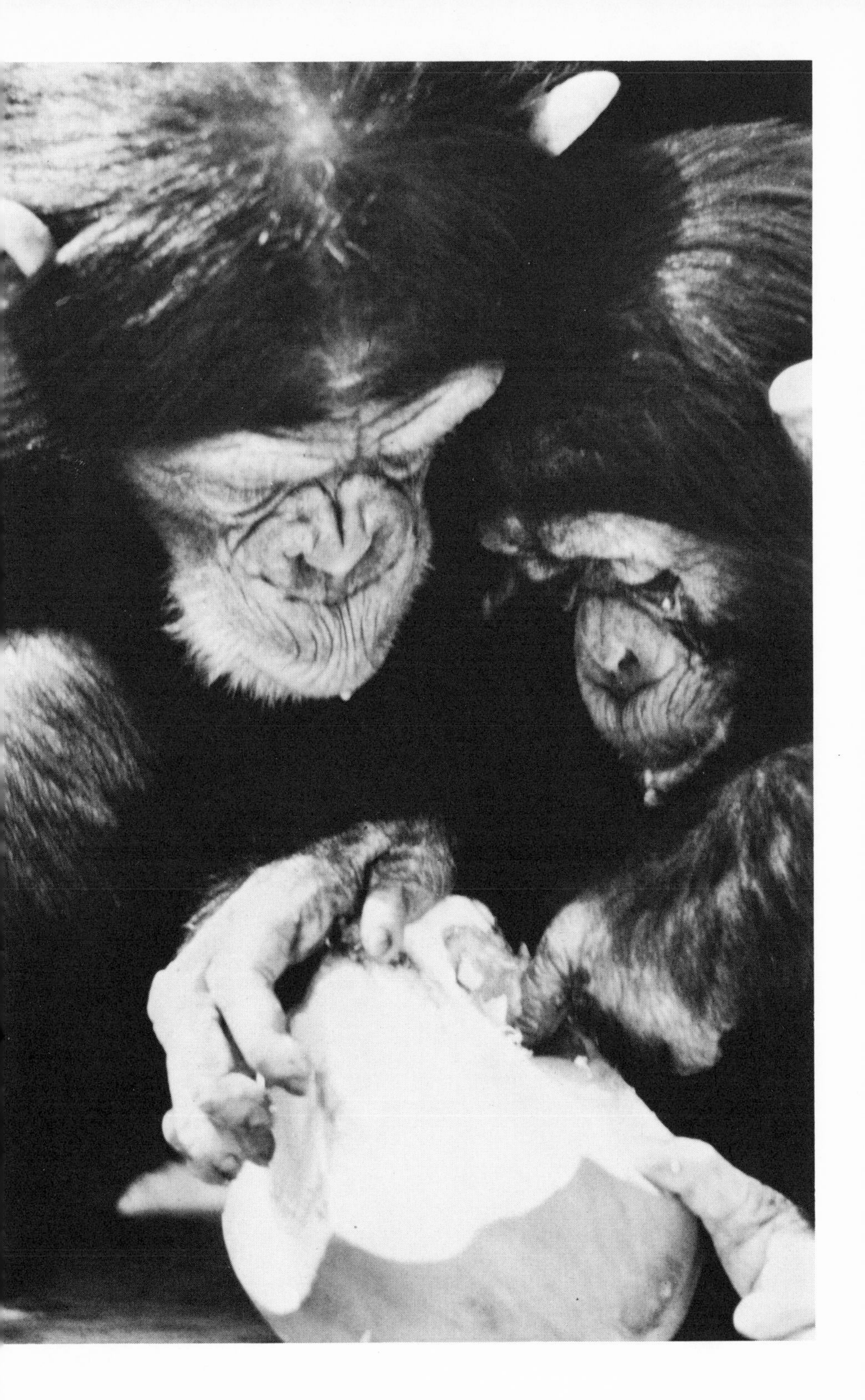

pretty chimp with a thick, long coat. The hair on his legs almost covered his feet and gave the impression that he was wearing a pair of bell-bottom trousers. He had a round, pale face with an enormous pair of soulful brown eyes and was probably about eighteen months old. He weighed 9–10 pounds and was 15 inches tall. He was terrified of humans and shunned any attention from us. We kept him in the house long enough for him to get used to drinking milk from a bottle, then took him to the reserve and presented him to Tina.

To our great relief the introduction worked like a dream. For Tina and Happy it was love at first sight, and Tina turned out to be an exceptionally attentive foster mother. She carried Happy around for short periods, and when she grew tired he walked beside her, clutching a handful of the hair on her back. Each evening when she came to the Orphanage for her supper, we were able to give Happy the milk which he still required, before he followed Tina into a tree and settled down next to her in the nest she made.

7. BACK TO ENGLAND

IT HAD BEEN WELL OVER A YEAR since I'd left school, and I was still entirely dependent upon my parents. I knew I was being useful in helping my father look after the reserve, but somehow, I felt, I had to earn a living. Prompted by a growing sense of restlessness, I wrote letters of application to the directors of two of England's largest animal parks—Longleat and the Woburn Wild Animal Kingdom. Both places replied, offering me interviews, and within a short time I was once again on my way to England. My feelings were mixed—I was sad and unsure about leaving, but excited at the prospect of being independent, in a new job among new people.

At Woburn I was interviewed by the general manager. He and his wife had previously lived in Uganda, and we chatted enthusiastically about our respective parts of Africa. The nervousness I had felt on entering his office quickly vanished. Later I was given a tour of the park, and was fascinated by the assortment of animals roaming around. One scene in particular struck me—a trio of tiny elephants being led across a field by a tall, loose-limbed young man.

I decided I liked the place, and when a job was offered me, accepted it eagerly.

For the first few weeks I and two other girls looked after a herd of giraffe and some zebra. Our job consisted of getting the animals out of their stables each morning and into a vast field called the picnic area, where we would feed them. While one of us kept an eye on the animals, the other two would muck out the stables and sweep the yard in front.

The picnic area was one of the two places in the park where visitors could get out of their cars to stretch their legs and enjoy a picnic lunch. By 10 a.m., when the park officially opened, the three

Stella looking after the giraffe

of us, armed with megaphones, would be in the center of the field. Two of us followed the giraffes about to make sure that none of the visitors approached them close enough to get kicked, and the third had the awful task of keeping up with the noisy, energetic zebras, for the same reason.

In the evening we would put the animals back in their stables and feed them again. The giraffes were usually quite docile, and the fact that they all slept in the same enormous shed cut down on the problems of getting them back in. The zebras, however, had to be put into five separate loose stalls. The difficulty was that only certain zebras could share a stable—otherwise there would be horrendous fights before morning came. Sorting *them* out for the night was often a fiasco.

During the long hours of giraffe-watching I'd look forward eagerly to the periods when the three baby elephants were in the picnic area. I'd go over to see them, or Tilly, their keeper, would bring them to stand nearby. They were African elephants, all under two years old. Tilly had looked after them since they had arrived in England, and they obviously adored him, following him about in much the same way they would their mothers, and forever trying to touch and nuzzle him with their rubberlike, mobile trunks.

Shortly after I arrived at Woburn, Tilly told me that he was being transferred to another park. I was sad at the news, and wondered who would look after his elephants. A day or two later Tilly informed me that *I* would be taking over his charges. I could scarcely believe it, and in my excitement little imagined how complicated "taking over" the young elephants was going to be.

First, there was a handing-over period of several weeks when I was with Tilly and the elephants all day long, being taught how to feed and care for them. While Tilly was with me, however, I might just as well not have existed as far as the elephants were concerned. The day finally came when I had to attempt to lead Tess, Dibby and Bunga out into the picnic area by myself.

With some foreboding I opened the stable door, and the elephants, one by one, stepped cautiously out into the yard. All three extended their trunks and flapped their ears, rumbling uneasily as they sought the familiar smell and sound of Tilly. It was very clear that if Tilly weren't there, they really would prefer to stay in the stable today, thank you very much. I did everything I knew to persuade them otherwise, but I received the distinct impression that they considered me a mild nuisance, and that my wheedling and begging were irritating, to say the least. Defeated, I ran to get a sack of their food and some of their favorite tidbits.

This time I bribed them out of the stable and quickly shut the door behind them. Still bribing furiously, I managed to get them past the office building in a reasonably orderly fashion, and with mounting relief passed through the main gate into the park. From here it was a matter of a five-minute walk to the picnic area. We'd covered about half the distance when I realized that my pockets were no longer bulging, and that I was, in fact, running very low on chopped carrots and apples. I tried to be less generous with my handouts, and the elephants accordingly became less interested in where I was going.

Suddenly it seemed to dawn on them that they had been tricked into walking much too far away from the security of the stable without their beloved Tilly. The gravity of the situation seemed to strike them all at the same moment, and they stopped. Tess wheeled around and began to run for all she was worth, with Dibby and Bunga close behind. High squeals of fear and excitement could be heard as the three wrinkled rumps, with tails held high, disappeared down the road. The sight would have been amusing had I not been so worried at my total helplessness to restore some calm and order. Hitching my bag of food to a more comfortable position, I hurried as fast as I could after them.

Tilly had been secretly watching our progress, and when I reached the main gate I was thankful to see that he had intercepted the flight and that the elephants were secure and relaxed again. The next time I took them out alone I borrowed Tilly's jacket, making sure I had a sufficient supply of tidbits and their sack of food. With these aids I managed to get to the picnic area, and my spell of success was prolonged while the elephants ate their breakfast. Then the search for Tilly began again. The more they searched, the more uneasy they became, until they were charging about nearly crazed. No amount of soothing words, commands, or anything else on my part helped in the slightest—in fact, it seemed I excited them even more. The only course of action was to radio an SOS to Tilly on the giraffe section's walkie-talkie.

Every evening I spent hours with the elephants, talking to them, feeding them and generally trying everything I could think of to win their confidence. Each morning I would set out in Tilly's baggy jacket, a sack of food over my shoulder, and three formerly exemplary elephants milling about uncertainly behind me.

As the weeks passed, the elephants came to rely on me in much the same way they had Tilly. They were winning creatures, each with its own distinct character. I became very fond of them—especially of the youngest, Bunga. She was intelligent and quick to learn the words and commands I began to teach them.

During the winter it was normally too cold for the picnic area, but I was allowed to take the elephants for a brisk walk in the adjoining estate grounds. I enjoyed these days the most—it was almost like being out with the chimps. I would catch the elephants' mood of exhilaration at being freed from the stable, and would run along beside them. It was fun to watch them play—flinging up dead branches with their trunks, only to back onto them and kick them

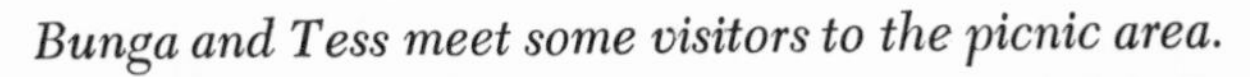

Bunga and Tess meet some visitors to the picnic area.

around with their feet. Sometimes, heads held high and ears flapping, they would make short charges into the bracken with shrill trumpets of excitement and tumble to their knees. We found a slope which they converted into a slide, careering uphill, then sliding down the muddy tracks almost on their bottoms. In the woods they would stretch their trunks to reach the low branches of an oak, and would then pop leaves daintily into their mouths.

Spring came round all too soon, bringing with it the picnic area routine. This time, however, the intimacy we had acquired over the winter months made the days among the crowds of visitors far less frustrating and difficult for both the elephants and myself. Also, a new keeper had been hired to look after the park's two older elephants—his name was Nigel Orbell, and we became good friends.

That summer the long days answering questions and repeatedly asking people not to feed or touch the elephants were interrupted occasionally for trips outside the park with the elephants. Once, Bunga and I took part in a carnival in Scotland, and afterward attended the Earl of Stirling's birthday party. Bunga behaved beautifully, and amused everyone by plucking flowers from the carnival queen's float and waving them in her trunk, or flinging them generously at the throngs that lined the street. At the birthday party she distinguished herself again. She walked into the elegant old house with the same easy familiarity with which she entered her stable. Without any hesitation she went straight to the table and began to help herself to cakes and tarts, raising mixed squeals of delight and apprehension from the seated children.

One morning I was given the news that an elephant of mine was to be transferred to a zoo, and that I was to choose which of them was the best suited to go. I'd often reminded myself that no matter how fond I was of the elephants, they did *not*, in any way, belong to me, and I was merely looking after them; still, it was a shock to realize that there was absolutely nothing I could do to prevent one of them from leaving. Even more distressing was the fact that I had to decide which one was to go to the lonely and depressing life of a permanent enclosure.

Stella and Bunga

For the first time since I'd begun work, I understood just how far from my ideals I'd strayed. The animals, the people, the way of life had been new and exciting, and I'd enjoyed it and learned a lot. But now a sense of perspective began to emerge, and I wondered why I was staying. Apart from keeping three young elephants content and healthy and answering questions about them all day, what was I achieving? Except for gaining some experience, very little. Why was

I working to help a rich place get richer, while my father struggled at home, and for so much worthier a cause? I felt a pang of shame that I'd ever considered leaving home at all. I deeply missed the animals there and the responsibility for their welfare. I wanted desperately to go home. I wrote and told my father so, and began saving what I could from my wages toward the trip back.

A week later, the day I'd been dreading arrived. A large horse box was backed into the yard. Steeling myself in a vain attempt to hide any emotion, I walked into the waiting box, and Dibby followed confidently. I rubbed her cheek and gave her a piece of apple. I said good-bye and tried to make her comfortable—which was only to wish her luck and as happy a future as people could give her.

After Dibby's departure Tess and Bunga were moved into the stable with the two older elephants. Nigel was able to bring the older animals into the picnic area, and my elephant-watching became less lonely. On the quieter days we had hours to talk, and I found myself telling Nigel in detail about The Gambia, Abuko and the chimps. Nigel had visited East Africa once, and had loved it. His ambition had been to return to one of the East African parks as a game warden, but he had found it virtually impossible to do this, and so had decided to try working at Woburn as a substitute. We talked about how ideal it would be if he could come to The Gambia and help at Abuko. It was just a dream, of course—there was, naturally, the wretched, insurmountable problem of money.

AMONG THE SMALL PILE OF MAIL awaiting me on my birthday that year was a card from Heather and my parents, and enclosed was a long, newsy letter. At the end of the letter was a P.S.: "It was difficult to decide what to give you for your birthday, but we all finally agreed to make an air ticket home available should you still want to come."

Should I want to!

The only real regrets I had were leaving Tess and Bunga and, of course, Nigel. The elephants were familiar with Nigel, however, and I knew that he would look after them for as long as he could. I'd worked at Woburn for just over a year when I gave my notice. I'd learned a lot—but I was extremely excited to be going home.

8. HOMECOMING

WHILE I WAS AWAY our group of chimps had grown to eight. William, Ann, Cheetah, Albert, Tina, Happy, and the two new arrivals during my absence, Pooh and Flint. Flint had adopted Ann and lived in the enclosure. Pooh lived free in the reserve as Tina's second adopted child. All the chimps had grown more than I had expected. Strangely enough it was not Ann who was the most demonstrative on my return, but Cheetah. He seemed beside himself with excitement and sat in my lap to groom my face and hair with frenzied enthusiasm.

I met Flint for the first time. He was an outgoing, independent chimp of about 3 years, and reminded me in many ways of William when he'd been younger. Ann was aloof, and though she hugged me sedately I felt that she did not clearly remember who I was. Now with a constant companion in Flint, she seemed perfectly content and self-contained. William was if anything more confident and roguish than ever, but grinned and hugged me affectionately. I hardly recognized Tina and Albert. They were seated on a low branch of a tree near the enclosure, and as Abduli approached with a tray of fruit, they food-grunted eagerly and descended, standing a few yards away as Abduli put their food on the feeding table. Neither of them paid any attention when Happy and Pooh were given their milk. It was clear that Happy was very reliant on Tina, for having finished his milk he went straight to sit beside her.

Pooh's behavior worried me. When he had finished his milk he immediately gathered a pile of leaves and sand into his groin and began to rock violently from side to side. I learned that Mummy had looked after him to begin with and had almost immediately dispensed with the diapers and frilly pants which must by then have seemed as much a part of him as his fingers and toes. Mummy said that he was conspicuously quiet in the house compared to William or the others. He spent most of the day rocking himself, and if Mummy found him making a puddle and picked him up to take him out onto the lawn, he would tense himself and go rigid, while his face remained unnaturally vacant.

Though his French foster parents had doted on him, they had preferred Pooh to act like a human child rather than a chimp. He was not allowed to climb trees, or do the normal things small chimps do. Consequently Pooh was disturbed and deprived. There seemed nothing physically wrong with him but he was definitely backward. It was thought that he might improve if he were with the other chimps, so he had been put into the enclosure a few weeks before I returned.

Stella and John Casey, an officer at the Reserve, with Flint, Pooh, William, and Cheetah

Pooh was an outcast from the first. William and Cheetah especially had bullied him. He was dragged about, bitten and allowed nowhere near the feeding table. He responded to this treatment either by rocking or going rigid. As a last resort, he was introduced to Tina. She was gentle with him, but Pooh hated all chimps and wanted to avoid them completely. My parents had felt that if we were too soft with him he would never become normal and accept the chimps, so much against his will he was left with Tina. She was extraordinarily patient with him, despite the fact that she had her hands full with Happy. Once, when Albert, Tina and Happy were nesting, Tina became concerned that Pooh had remained rocking in the grass at the base of her tree. She went down and tried to coax him up, but to no avail. Finally she took a firm grip on his upper arm and dragged him up into the tree. The moment she released him he rewarded her efforts by half-tumbling, half-climbing down to the bottom of the tree again and rocking frantically in the grass.

The day following my return I automatically took over the chimps again. I felt sorry for Pooh—he was such a lonely figure in the group. He seemed to sense my sympathy immediately and tried to climb into my arms. At first I carried him or allowed him to sit on my lap and hug himself to me. I sensed, rather than knew at the time, that the cure to Pooh's neurosis was affection and security.

Pooh began to lean on me more and more, and while we were out on the walks he would always remain close to me. He was a finicky feeder and therefore always rather thin and scrawny, but gradually, with new confidence, he started to improve. He took a more lively interest in what was going on around him, and if I played with him or tickled him I would be rewarded with a few husky pants of laughter and the open-mouthed play face common to all chimps. The game "this little piggy" was one of his favorites, and even after the little piggy had gone "wee wee wee all the way home" for the twentieth time, Pooh would still stick his pink toes in my face for more. I would get tired of the game before he did and alternate play with grooming. The moment I began to part the hair and search his coat I could feel him relax. And if I groomed him long enough he would drift off into a light sleep.

His attachment to me and my constant care of him had the effect of minimizing William and Cheetah's bullying—also, I think, the fact that he now had more self-confidence and occasionally showed some spirit helped in this respect. If he was bullied or left out I would pick him up and try to distract him from his rocking with play or some other attention. It seemed that having someone to run to, to protect and comfort him, made a world of difference to him. He began to imitate the other chimps and joined in their games. When I left the reserve at the end of the morning Pooh would follow Tina, Happy and Albert. He was also learning to feed on the different fruits and leaves that grew in the reserve.

It was exciting and interesting to have a group of eight chimps to walk each day, rather than the three that I'd had before. There was not an undisputed leader of the group, though Cheetah, due to his size and character, came closer to claiming the title than anyone. All the chimps except William and Tina were afraid of Cheetah and showed him deference, but if he became over-aggressive two or three of them would gang up and retaliate. Albert, who by now was a regular visitor, was the most afraid of Cheetah. Their small disputes were never serious, though, and a few minutes after a squabble it was common to find the same individuals playing, feeding or

Pooh, age five

Daddy with Tina (left) and Pooh (right)

grooming together. None of the chimps had set play partners. William and Cheetah were close companions and showed mutual respect for one another, but William played just as frequently with the little ones or with Tina as he did with Cheetah.

There were no predators in Abuko that the chimps needed to fear. The only real danger was snakes, and the chimps usually showed a healthy fear of these. Sometimes while following the winding path that runs through the reserve, the leading chimps would round a corner, then all suddenly reappear, either climbing trees or gripping some part of my anatomy for protection. If I crept forward quietly I would usually find a snake stretched across the path or coiled in a branch close by. On these occasions I feigned exaggerated fear and retreated too. We would either make a large detour through the bush or wait until the snake moved off.

William was the only chimp in the group who showed any tendency to attack snakes. During one peaceful walk Tina suddenly screamed and leaped a good few feet up the trunk of a nearby oil palm tree. From there she repeatedly made the loud whaa barks which I'd learned meant there was something below her she considered alarming. I walked cautiously in the grass but could not at first see what was causing Tina's distress. Then I heard a powerful and seemingly continuous exhalation of air. Just ahead of me, stretched along a fallen tree trunk, was about twelve feet of African python.

As I began to retreat, William suddenly charged into view with the base of a dead palm frond in his hand. This he flung with all his might at the python, whaaing as he did so. There was a chorus of excited calls from the rest of the chimps. I picked up William and carried him till I considered we were a safe distance away, and the other chimps followed. Very rarely had I seen any of the chimps display overt aggression toward another animal in the reserve. It was clear that William's growing self-confidence and security in his surroundings were going to present more and more problems as he grew older.

9. PROBLEMS AT THE RESERVE

AS THE WEEKS PASSED I began to notice that Cheetah was showing an increasing amount of interest in Tina. It didn't take me long to realize that Tina—now 8 years old—had reached puberty and that the small pink swellings which periodically appeared around the ano-genital area made her extremely attractive. Cheetah became possessive and jealous. Perhaps William preferred not to interfere with what Cheetah considered his, for he showed remarkably little interest in Tina and was quite content to hug Happy, or allow him to walk alongside clutching a handful of hair on his back.

As Tina's sexual swellings developed, so did Cheetah's possessiveness. Tina did not seem to mind being protected, for Cheetah was her favorite suitor and she would follow him without hesitation when he led her a little way off from the main group. I noticed sadly that when she was in this condition she was far less attentive and protective toward Happy, so Happy spent more and more time playing with Pooh, Flint and Ann or being cared for by William and Albert.

At first Albert resented the time and attention Cheetah lavished on Tina. When Cheetah went back into the enclosure, Albert—being the only available male of any size—would take his place. Albert was still afraid of Cheetah, but would try every tactic he knew to distract the other chimp's attention, sometimes shaking branches at him, then running away so Cheetah would chase him—though he knew that if he were caught the results would be painful and degrading. He kept on for a long time but finally gave up when Tina joined Cheetah and both of them chased Albert away. Albert had to accept the fact that when Cheetah was around he no longer had access to Tina.

Frequently when Cheetah mated Tina, Happy, Pooh and Flint would become very excited and rush over to the pair. Cheetah was tolerant to a degree, but if they persisted too long he would chase them off. Ann would interrupt her perpetual feeding to follow Tina and Cheetah to their chosen recess, and though she sat at a respectful distance and did little more than stare, she too was driven away. When Tina's swelling disappeared the group reverted to what it had been, except that everyone now paid slightly more attention to Cheetah.

ALMOST EVERY DAY while out with the chimps, we would meet colobus or green vervet monkeys. These were not excessively timid and would often feed in nearby trees. I remembered that when there had been only Cheetah, William and Ann, the chimps had actually fed in the same trees as the colobus, and though Ann would sometimes stamp her feet on the branch if she found herself too close to

Vervet monkeys near the crocodile pool at Abuko

one, the chimps had shown no real desire to chase them off or include them in play.

Since my return I noticed that colobus-chasing was becoming an increasingly popular game. The colobus still didn't seem unduly perturbed and would allow the chimps to get quite close before tripping off confidently through the leaf canopy or leaping easily the enormous distances into a neighboring tree.

One morning Tina, Cheetah and Albert were in the lead, William followed with Happy, while Flint, Ann and Pooh were walking beside me. I saw Tina suddenly leap forward and run into some dense undergrowth, with Cheetah and Albert close behind. There was much rustling of dry leaves, and I thought the chimps were chasing a pack of mongooses. The sudden high squeaking seemed to confirm this, till I heard an outburst of loud, excited screaming coming from the hidden chimps.

William raced forward and the other chimps followed. As I crawled through the tangle of thick vegetation to where they were, I thought they were having an almighty fight; everyone was screaming hysterically and seeming to dive in and bite or pull at Tina's stomach. As I approached, Tina leaped clear and ran, still screaming, up a small tree. All the others were still very excited and several of them embraced each other, then hooted and barked again with renewed vigor. I couldn't understand the frenzy till I saw that Tina held in one of her hands a very young green vervet monkey that was still alive and squeaking. I stood paralyzed for a moment, then ran to Tina's tree. This caused a renewed explosion of noise from the rest of the chimps, and Cheetah began to climb toward Tina.

I asked her in a shaky voice to give me the baby monkey and held out my hand. The screams and whaas grew louder. I couldn't understand why, but clearly they were intent on harming the baby monkey. I had raised and looked after several vervets and found them loving little creatures; that this mob should want to hurt one so small and helpless was unbearable. I *had* to rescue that infant. I ordered Tina again to drop the monkey and flung a small branch at her—she climbed higher, still uncontrollably excited, then brought the monkey to her mouth and bit into its head ripping off an ear. The squeaking stopped and Tina began to bite and eat.

I couldn't watch. Feeling thoroughly sick, I fled back to the path.

The young chimps followed but I wanted nothing to do with them. Even Pooh had been trying to grab and bite at the monkey, and when he took my hand it was all I could do to let him hold it.

I was shocked and angry. Where would you all be, I cried to them, if, when you had arrived on the doorstep as helpless babies, I had picked you up and bitten your heads off? Very upset, I returned to the Orphanage. The chimps caught up with me on the path, and I turned once to see Tina—watched closely by the others—sitting eating a leg of her prey.

On reaching the Orphanage, I sat in the rest house to think. Tina appeared, but without the monkey, and sat close to the wire grooming Cheetah. Looking at them all again, it was impossible to believe how savagely they had been behaving half an hour earlier. I had supposed that chimps were strictly vegetarian, and Tina's killing and eating of the small vervet seemed to me cruel and degenerate. But having thought about it, I came to the conclusion that they were still my chimps. If they were disturbed and deranged as a result of their capture and present life, I had to try even harder to help them get over it.

I shall always be grateful to my friend Jan Morrison for helping to shed new light on the whole episode. She came to the reserve one evening and I told her of the monkey incident. To my surprise she didn't say "Poor little monkey" but clapped her hands with delight. She went on to explain that she had just read a new book, *In the Shadow of Man,* by a woman named Jane Goodall who had been accepted by and lived among a group of wild chimps in Tanzania. She was making a study of their lifestyle and behavior. One of the most significant things she had discovered was that wild chimps hunt small mammals in order to eat them. Jan promised to bring me the book the following day.

The book was a revelation. I read it through several times and studied the photographs for hours. Here was a detailed and vivid account of the day-to-day lives of a group of wild chimpanzees. I could picture clearly the sort of background from which our chimps had come and more than ever wondered what their families had been like. It made me realize that though we had been able to give our chimps a far more natural lifestyle than most such captives, they were still quite deprived.

After reading *In the Shadow of Man,* I came to view Tina's hunting of the baby monkey differently. I now saw it as a compliment to us that she was still, despite the trauma of her capture and confinement, able to follow the habits of her species. Far from being deranged, she was teaching the other chimps the normal habits of a wild chimpanzee. Goodall's book helped me to understand with new clarity the behavior of my group of chimps. It no longer seemed strange to me that eighteen-month-old Happy was able to take such an interest in Tina's sexual swellings when I read of babies half his age doing the same. Compared to some of the females in the wild group, Tina's persistent invitations to be mated when she was receptive no longer seemed excessive or abnormal.

I understood why Albert crouched panting and presented his backside to me or the other chimps when he met us: it was a common gesture of respect and submission. Most of all I gained insight into Pooh's problem, for his behavior resembled closely that of a young, orphaned wild chimpanzee. With relief I realized that I had given him the best medicine for his problem—attention, affection and security—and I understood why he had responded so well and so quickly.

As a result of reading Jane Goodall's book I began to keep a diary in which I recorded all the major events of the morning walks and other aspects of the chimps' behavior that I considered important. With a written record to refer to, it was easier to follow the gradual changes in each individual and in the group as a whole.

During the ensuing year the younger chimps became increasingly more independent while out in the reserve. Though they still followed me closely, they showed less tendency to become distressed and frightened when they lost sight of me. They also became more confident and playful with people they did not know, instead of clinging shyly to me and whimpering if anyone else picked them up. Eventually they were actually leaving me to follow tourists down the path.

William and Cheetah were the most confident members of the group. They engaged every cooperative tourist they met in vigorous play, climbing onto their shoulders and tumbling down into laps or

William about to explore the contents of a purse

waiting arms. Another favorite game was to dangle in a relaxed position from people's arms, hoping to be swung. Most people loved the close contact with the chimps and treated them as exuberant young children, and were unusually careful and gentle with them during the games. There were exceptions, of course, people who tended to play too roughly, but I was always present during the walks to ensure that chimps and visitors did not break the rules of the game and become overexcited.

There was, however, a small percentage of visitors to the reserve who were terrified or even disgusted by the chimps. Chimps are extremely sensitive to human reactions—when they detect the slightest nervousness in anyone, their behavior changes noticeably. Our chimps would continually harass the person, playing far more boisterously than normal and seeming to take delight in the increasing fear and nervousness they were able to provoke. In these situations I distracted the chimps and advised the person to move on down the path as quickly as possible.

During the years the chimps spent at Abuko they came into contact with hundreds of visitors, many of whom were children, but there were only four cases of their actually biting a visitor, and in all instances it was the human's fault. In three of the four cases the visitor had no intention of provoking the chimps, but did so through not being able to understand their behavior. One such instance occurred when someone gave Happy a sweet. I did not like people to feed the chimps, but occasionally they were given sweets without my knowledge. At that time Cheetah saw the visitor slip a sweet to Happy. He ran over to her and asked for one too. She gave Cheetah a sweet and took the opportunity to place a second one in Happy's hand. Cheetah immediately reached over and took the sweet from Happy. The woman, incensed at Cheetah's unfairness, snatched it and tried to give it back to Happy. Cheetah flew into a rage, bit the woman's hand, then attacked Happy. It was only when the screaming started that I realized what had happened. It took me a long time to explain to the woman that she had almost asked to be bitten.

Although the chimps were rarely aggressive toward visitors, they were quick to realize that they could take more liberties with them than with me. Visitors simply did not know what to expect from the chimps and therefore accepted everything. The chimps soon learned this and took delight in misbehaving when they knew they could get away with it. They had been taught that it was wrong to steal objects from a pocket or purse. I often took a shoulder bag out on the morning walks, and the chimps understood that to remove anything from it was forbidden. If I caught them trying they could expect punishment in the form of a scolding or a slap. The younger chimps considered slapping a very severe form of punishment, not because it hurt—for they often slapped each other just as hard while they played—but because of the implication of rejection it conveyed. After a smack, the chimps whimpered or screamed, then held out their arms to be reassured of my affection.

While there were still relatively few visitors to the reserve it was easy to ensure that the chimps continued to respect the rule about stealing, but as the number of visitors increased it became more difficult to watch all eight of the chimps at once. William and Cheetah began to snatch people's hats and refuse to return them. Most visitors found this amusing, especially when Cheetah would chase William through the branches of a tree and attempt to take the hat from William's head. The more applause the chimps received, the more excited they became, until they were deaf to my requests to return the stolen article.

The chimps were able to sense when I was merely acting and when I was really angry. I usually managed to retrieve the hats only when William and Cheetah's blatant indifference to my orders annoyed me to such an extent that I felt genuine fury. At that point Cheetah and William would reluctantly throw down their loot.

William and Abduli

Each time the chimps stole hats, though, they became increasingly immune to scoldings. William and Cheetah realized that if they remained in a tree out of my reach, there was little I could do except lecture them. The result was that they formed a team, and from then on the fun of meeting tourists lay in how much they could steal and how quickly they could race up trees and get their trophies out of my reach.

The theft of hats and pens was harmless enough, but when the chimps became more ambitious and began to steal spectacles and bags, I found I was facing a serious problem. William was incredibly cunning when it came to acquiring such articles. While he enthralled his victims with his games and antics, he would be angling to get into a position where he could slip his hand into a pocket and seize whatever it contained. The moment he had something—usually a wallet or a comb—in his grasp, he would leap away and scramble up a tree. Cheetah would follow and they would chase each other through the branches, playfully fighting for possession of whatever had been stolen.

Despite my warnings to the visitors to hold on tightly to their belongings, people would often respond to William's polite requests to peep into a basket or handbag. As his backside vanished into a thicket with their purses or sandwiches, they would look amazed that such a charming little fellow could change so quickly into a thieving rogue. I began to avoid the path when I took the chimps out. However, the reserve is very small, and from the top of a tree William could sometimes see the path and anyone who happened to be using it. On these occasions he would quietly slip away, and it was often several minutes before I noticed his absence.

I remember once hurrying off in pursuit of William, to discover an overweight Swedish lady crawling on her hands and knees beside the path. Looking somewhat disheveled, she straightened up and smiled self-consciously. She could speak very little English, but I had already guessed what had happened. At the far end of a tunnel through a particularly dense patch of vegetation I could hear William's squeaks of recognition. As I crawled along I found a trail of objects: bank notes, address book, cigarettes, a lighter, a broken compact, a semi-eaten lipstick—and finally William with an empty handbag beside him. He had a small comb stuck behind his ear and was busy playing with the zipper of a purse. As I approached he moved farther down the tunnel, and I realized that I would have to use more subtle tactics if I was to recover the rest of what he had stolen.

In my shoulder bag I kept a small supply of sweets for such emergencies. I offered William a sweet with one hand while holding my other hand out toward him. He was interested in the sweet and offered the equivalent of a twopenny piece for it. I shook my head and pointed at the purse. William understood and to my relief picked up the purse, but instead of giving it to me, he opened it and tipped the contents out in front of him. I handed him a sweet, amazed at his good business sense. Six sweets and much bargaining later I had retrieved almost all that William had taken and was able to wriggle back to where the Swedish lady waited. Pooh was sitting on her shoulders, carefully taking the hairpins out of her bun.

I HAD HOPED THAT when the summer rains brought an end to the tourist season, the morning walks with the chimps would once again become reasonably peaceful. However, the introduction of Charlie Oribi into the reserve produced a new kind of chaos.

An oribi is a gray-brown antelope about the height of an average dog. On either cheek it has a bald black spot, and beneath its eyes are small glands which exude a liquid it rubs on vegetation for scenting purposes. Charlie was a young male oribi with short, blunted horns. When his owner brought him to the reserve and presented him to us, he was very tame and looked a healthy, handsome young creature as he rubbed his glands on his owner's leg.

Rather than put him straight into the main reserve, we thought we would leave him in the antelope enclosure for a few days to let him settle down. Several hours later he was chasing the other antelopes around so persistently that I was sent into the enclosure to try to coax him through the gate again. Charlie was in a state of high excitement, careering round the large enclosure, leaping high in the air or making stiff-legged hops over the ground when he tired of sprinting. Every so often he would pause, panting, just long enough to rub his glands on a blade of grass or a twig. As soon as I entered the enclosure he bounded over and began circling my legs.

All the while he kept up a pathetic gurgling, bleating sound. It sounded so plaintive that I reached down and tried to comfort him, thinking he was missing his owner. He submitted to the caressing for a moment or two, then dashed off, spun round and came galloping toward me with his head down. I had just enough time to sidestep to avoid his horns. I quickly led him toward the gate. He followed for a short distance, then began to try to butt my legs again. I managed to catch his horns before they made contact, and dragged him unceremoniously out into the main reserve. The rough treatment frightened him, for he cantered off into the bush, making the high-pitched whistling sound which is an oribi's way of expressing alarm.

Occasionally Charlie would meet us when out on a walk; sometimes he would follow us peacefully for a short distance and then wander off alone, at other times he could be a real pest. If there were any wild female oribis living in the reserve, Charlie had failed to find them, and I was convinced that frustration was the root of his particular problem. All the chimps except William were afraid of him, and he disrupted many a quiet walk.

One day we were out for a walk when an all-too-familiar bleating and a sudden rustle of leaves made me turn my head quickly. Charlie Oribi stood a few yards behind me, rubbing his horns in an irritated fashion on the branches of a low bush. I remained still, and optimistically waited for him to go away. He might have done so had William not swung down from a nearby tree to tease him. A scene was then enacted very similar to that found in a Spanish bullring. With all the arrogance of a true matador, William baited his bull, throwing dust in his face, flinging dead branches, and snatching at his back legs before leaping to the safety of a tree. He avoided Charlie's charges with admirable bravado, slapping the infuriated oribi heartily on the rump each time he passed. He then retired, panting, to a tauntingly low branch from which he continued to tease the irate antelope.

Charlie looked around for easier revenge. Stiff-legged with thwarted rage, he turned to face me. I fled, and would have climbed a tree had Pooh not mistaken my flight for desertion. Screaming, he leaped from a branch, landing on my shoulders with a thud. I faltered beneath Pooh's sudden weight, which gave Charlie the advantage he needed. As I grasped the low branch of a tree, I felt the impact of his horn on the back of my leg.

William descended, perhaps to come to my rescue, but as he approached, Charlie spun around in a spray of damp soil and charged straight at him. William dodged an instant too late and received a glancing blow on the chest from Charlie's shoulder which sent him sprawling backward. He was struggling to his feet when Charlie came in again with lowered head and hit William full in the face. There was an agonized shriek. I raced to William, Pooh still clinging to me, and saw both of William's hands clasped tightly over his face. Charlie, who had galloped around in a frenzied circle, was bearing down on us once more. Grabbing the nearest thing to hand, a heavy piece of dead wood, I hurled it at him. It struck him on the flank with sufficient force to make him swerve and sprint away into the scrub, whistling his alarm.

I bent down to examine William. Blood was oozing from between his fingers and slowly trickling down over the dark hairs of his wrist. Firmly but gently, I prized his reluctant hands away from his face. His left eye was a swollen, gory mess. Gulping down rising panic, I picked him up. Pooh refused to walk when I pushed him off my back, and clutched my leg, screaming. I realized I would have to carry them both. As I reached down for Pooh, I noticed a patch of red flesh just below the back of my knee. Surprisingly, I could feel no pain. Calling to the other chimps, I hurried back along the path toward the Orphanage.

We rushed William off to the vet. The injured chimp whimpered and clutched at me when I tried to seat him on the table, but from then on he behaved perfectly. In a few minutes his eye had been anesthetized and the vet was lifting his face with a large freckled hand and inspecting the swollen eye area. I thought I would be told that an urgent operation was required to remove the rest of his eye or, worse, that William was going to go blind. It took an effort to remain calm and sensible.

Finally came the verdict. "He's very fortunate," I heard the vet say. "It's not so serious as it looks. The eye itself is miraculously undamaged. Seems that the horn of the beast passed in front of the eyeball, through the eyelid, and was stopped by the chimp's prominent brow."

We decided against a stitch in the split eyelid, as it would have been almost impossible to prevent William from picking at it and perhaps causing more damage. The vet gave me an assortment of medicines and walked to the Land Rover. I shook his hand warmly; William grunted and halfheartedly extended his hand too, which the vet smilingly accepted.

As soon as we reached the house, I gave William some fruit juice, then laid him on the couch. Heather, who had been in the middle of her lunch, hurried into the bathroom to prepare a dressing for my leg. William spent the rest of that day curled on the couch with a pillow pressed to his face. Duff's exuberant approaches were met with short-tempered rebuffs, but William seemed to welcome Tess, who was frequently found lying either on the carpet beside the couch or, if she thought no one was looking, actually on the cushions next to him.

William was touchingly trusting when I needed to bathe or treat his eye. He would even hold his face up, keeping absolutely still. He recovered quickly, vacating his couch more and more frequently to begin a rapid destruction of the house. After an exceptionally exasperating day of rescuing household objects, turning off running water and cleaning up puddles, we decided that enough was enough, and William returned in triumph to the enclosure.

10. THE LAST STRAW

THE TOURIST SEASON HAVING ENDED, the chimps replaced the amusing distraction created by the visitors with more frequent monkey hunts. Tina was the hunt leader, with Cheetah, Albert and William her close assistants.

To begin with, chasing the colobus was a haphazard affair. Although all the chimps joined in, I could see little cooperation among them; it was more a matter of each chimp for itself. Success was a matter of luck rather than the result of strategic hunting. The more practice they had, though, the more organized they seemed to become. I once watched Tina chase a colobus until it made a tremendous leap into the crown of an isolated oil palm. Cheetah was close to Tina and remained on the branch from which the colobus had leaped while Tina backtracked and scrambled quickly to the ground. This blocked the only possible aerial escape route. Albert was already at the base of the oil palm when Tina arrived, but kept sitting there and allowed Tina to climb. William stood a few yards behind Albert and watched Tina's progress. As she neared her prey all the chimps began to whaa excitedly, yet remained in more or less the same position.

Tina quickly closed the distance between herself and the terrified monkey, but at the last moment the colobus leaped to the tip of a palm frond, which sank beneath his weight, then made a thirty-foot drop into a tangle of low bushes. Albert and William were waiting, however, and between them they caught the colobus. Tina, Cheetah and the four younger chimps soon joined them. It was still horrifying for me to watch them kill, but I had made up my mind to interfere only in exceptional cases.

Among the eight, only Tina and Albert actually ate the meat, which they would chew with green leaves exactly as had the community of chimps in Tanzania described by Jane Goodall. I hoped that through watching Tina and Albert the others would learn to make proper use of their prey rather than turning it into a macabre plaything. They would watch Tina and Albert intently as they ate, pushing their faces to within inches of the chewing mouths, and Ann and William especially would constantly sniff and occasionally

Albert eating a vervet monkey

taste small pieces of fur and meat, although neither really ate anything. When Albert and Tina had eaten their fill, the others would treat the colobus as a toy, pulling it away from each other, rolling around with it or wrestling for possession of the gory remains. Once the colobus began to smell or attract flies, however, the chimps would abandon it and thereafter carefully avoid going near it.

Altogether during those wet months the chimps killed seven colobus infants and one sub-adult female. They would also have killed an adult female had I not intervened and rescued her.

Colobus were not the only animals the chimps chased. They were also successful in catching two vervet infants, and would frequently chase squirrels and other small mammals. Once I was appalled to see William emerge from the undergrowth with a freshly killed vine snake slung round his neck. I was never able to find out whether he had killed it himself or had merely found it. Also in the enclosure I frequently found dead mice, lizards, toads and occasionally birds, all obviously killed by the chimps.

One evening some friends came to the reserve to feed the animals. Their small daughter, Clare, was with them, carrying her constant companion, a very lifelike doll called Cindy. All the chimps were good with children. William especially seemed genuinely fond of them. He would frequently hold hands with them and play far more gently than he did with adults, so I had no fear of bringing Clare with us.

Tina was late arriving at the Orphanage that evening, and by the time she appeared, Clare and her doll had already met the others and Clare had lost all her initial apprehension. The first warning I had that all was not well was when I saw that Tina had erected her coat and stood glaring at the doll in Clare's hand. Tina then approached and tried to take the doll from Clare. Clare took a step backward and protectively clutched her "baby" to her breast. Fortunately I arrived at this point, picked up Clare and her doll with as casual an air as I could manage, and carried them out of the enclosure. Behind me I could hear Tina whimpering. Then the whimper increased in intensity until she was screaming in the throes of a temper tantrum.

The incident really worried me. I never found out exactly why Tina was so desperate to have the doll. Her approach was quite different from that used in catching young colobus prey, yet I had never seen her so interested in acquiring a mere plaything before. I suspected that had I not been present, and had Clare insisted on keeping her doll, Tina might well have taken it by force and perhaps hurt the child.

It was a disquieting thought, and when it was added to the difficulty of walking the chimps in the tourist season, I began to realize that we were facing what could quickly become a major problem.

ANOTHER LESSER BUT STILL DIFFICULT problem soon surfaced. One morning when I got to the enclosure, I found it empty. The normally neat rest house looked as if a tornado had hit it. There were lumps of straw spread all over the ground and the sad, wilted heads of bougainvillaea flowers littered the yard. The toilets had been invaded and yards of toilet paper lay strewn all over the gardens.

I knew that Daddy would be arriving within an hour, so Abduli and I set to work like fiends to clear up the mess and make the place look presentable. I thought we'd done a rather good job of patching everything up, but the despairing look of incredulity that spread over Daddy's face as he entered the Orphanage was painful to see. "What's been going on?" he asked in a higher-than-normal voice. "What in God's name's been going on?" On examining the enclosure I could find no possible place where the chimps could have escaped. The doors were all still locked and there were no holes or other clues.

I found the chimps somewhere near the middle of the reserve. They seemed pleased to have me join them. Pooh was the most demonstrative and sat on my lap or near me for the best part of the morning. Their mysterious escape puzzled me, and it was not until a few days later that I witnessed what could conceivably have been the means of their release. In the enclosure we had built a triangular climbing apparatus from young gmelina tree thinnings held firmly in place with six-inch nails. I watched Cheetah working for about fifteen minutes to remove one of the climbing bars, which had obviously become slightly loose with use. Finally he was able to pull the bar away from the main structure. The six-inch nail, still reasonably straight, protruded from both ends of the bar.

Cheetah and William played with this bar for about ten minutes before Cheetah took it to the enclosure fence and began to climb the wire, dragging the bar after him. When he reached the barrier of corrugated iron sheets, he pulled the bar up and maneuvered it until the six-inch nail was hooked over the top of the enclosure. The bar then served as a bridge across the previously unscalable sheets of iron. Cheetah began to climb the bar to freedom, and William, now aware of what was happening, raced up the wire to join him. Unfortunately for Cheetah, the nail slid out of position as he climbed, and both he and the bar fell to the ground. Undiscouraged, he immediately climbed back, dragging the bar behind him, to where William still waited. This time, once Cheetah had hooked the nail over the corrugated iron, William hung from the free end of the bar, thus making it more secure, and Cheetah raced up and out. He was, however, in such a hurry that he did not hold the bar for William, who, though he tried, was not able to escape alone. I could hardly imagine William holding the bar in place while Ann and Flint climbed it, or either of them holding it for William. The entire procedure had surprised me so much that I was willing to accept almost anything; on the whole, however, I concluded that I had just watched an entirely new escape method and not the one by which they had all fled a few days previously.

The climbing apparatus was repaired, and for the next week none of the chimps left the enclosure without a chaperone. Then one morning I arrived to find the enclosure empty again and the Orphanage in total chaos. All the apparatus was intact, and as before I could find no clue as to how everyone had escaped.

Once more Abduli and I tried desperately to restore order before my father arrived, but it was impossible to repair all the destruction the chimps had wrought. Worst of all, Daddy's prized western sitatunga's enclosure had been opened and the young antelope had disappeared. This time my father was too depressed to be really angry. Sadly he turned to me and said: "The chimps are going to have to go somewhere else, you know, Stella. They are getting too much of a handful for us to keep in Abuko." The same thought had

Pooh takes a leap (William is at left).

been milling vaguely in my own mind for some time, but it had been so unpleasant that I'd pushed it away and concentrated on the day-to-day routine instead.

Somehow I had to find out how the chimps were escaping. For a week I arrived at the reserve before dawn and waited, but nothing happened. For another week or so all went peacefully and the question of transferring the chimps somewhere else was shelved again.

Then the time came round for Tina's swelling, and once more she and Cheetah were inseparable. Tina spent much more time in the Orphanage, and it was not uncommon to see Cheetah trying to mate her through the enclosure mesh. One morning when I arrived, Abduli rushed down the path to meet me and told me he had found out how all the chimps escaped. Tina climbed up the outside of the enclosure, which was easy since all the supports were on the outside. She then gripped the top of the corrugated iron and hung down on the inside, bridging the barrier. Cheetah ran up the wire, jumped to catch Tina's feet and climbed up her body to the top of the fence. Abduli had prevented the others from escaping in the same fashion, but Cheetah, he said, had run off with Tina.

All that morning I thought about how we could prevent Tina from releasing the chimps, but I could not come up with any satisfactory answer. To modify the enclosure would cost more than we could afford, and to enclose Tina would be more or less impossible. She might enter the enclosure once, but having found herself locked in I was sure she would never consent to do so a second time. I knew that it was only a matter of time before the chimps would escape again and create another series of disasters. Tina, Cheetah and Albert had grown fast during the last two years and were continually startling me with their tremendous feats of strength. Cheetah, who was always rather muscular, began to destroy the Orphanage gates, and eventually succeeded in ripping them right off their hinges. The stout mahogany benches placed at intervals along the path were frequently overturned, and most alarming of all, when we returned to the Orphanage after a walk one day, he played truant, refused to re-enter the enclosure, and tore the padlock off the hyena enclosure, almost letting the two hyenas out into the reserve.

A few days later I arrived to find that the chimps had escaped again. Pooh, Ann and Flint were having the time of their lives on the rest-house roof, pulling out tufts of thatching and scattering it around, rolling and playing with each other as they did so. Daddy had gone to great pains to find and plant a row of attractive indigenous trees along the front of the rest-house compound. Three of these had been destroyed and another partially uprooted. Cheetah, Tina and Albert were out of sight, but some of the chimps had obviously been playing in the young antelope enclosure, for most of the thatching on the two shelters we'd built there lay in a messy heap on the ground.

Happy, I discovered, had caught our tame vulture. The bird appeared dead, and Happy sat solemnly beside it, methodically plucking one feather at a time from its back. (The bird was in fact merely playing dead, and apart from being frightened and having a small bald patch on his back was perfectly healthy.)

Things could not be allowed to go on like this indefinitely. I had sworn to myself that whatever happened I would rather die than allow any of the chimps to be sent to a zoo. The only alternative was to release them somewhere else. I had not heard of anyone trying to do this before and wondered just how feasible it would be. I spent much of my time planning how I would go about returning the chimps to the wild and where I would do it if, or as now seemed certain, when the time came.

11. NIOKOLO KOBA

My father had visited a large national park called Niokolo Koba in Senegal the year before and had thought it a beautiful place. He told me of a small mountain there, called Asserik, to which he had driven in the hope of seeing or hearing the wild chimpanzees that were sometimes observed there.

Niokolo Koba National Park, just over four hundred miles away by road, would be the obvious place to release the chimps. For a second I allowed my imagination to wander and envisaged all eight of the Abuko chimps climbing, feeding and playing with a group of friendly wild chimps. Perhaps William or Cheetah would one day become dominant males, Tina would have children, Albert would no longer be frustrated and everyone would live happily ever after. But I knew many difficulties and hazards would have to be surmounted before that dream could come true.

Perhaps the chimps would not be able to survive alone in a strange place. Tina and Albert would, I was sure, manage, and perhaps they would teach and look after the others, but I would have to stay with the group until they became familiar with the new area, and possibly with new foods. How was I to do this? I didn't possess any equipment, even a tent, and having spent the last two years working voluntarily, had no money of my own.

I tried to calculate how much it would cost. Even being as economical as I could the sum still seemed vast and the whole scheme an impossibility.

I decided to begin at the beginning. First I had to get permission from the park authorities to carry out the experiment. Mr. A. R. Dupuy was the Director of National Parks in Senegal, and his help and approval would be essential. I wrote my first letter to him, and my second to Jane Goodall, asking if she could help in any way.

I got Mr. Dupuy's reply first, giving permission to release the chimps in Niokolo Koba and to remain with them to make observations on their release and post-release progress. However, I would have to stay at an established guard camp in the park, as the roads to Mount Asserick were flooded during the rains and the area was cut off. The park could not take the risk of allowing me to remain alone there under rainy-season conditions.

Jane Goodall's reply arrived a few days later. It was a warm, informal letter, saying she had hoped to send one of her students to help me but this had become impossible; however, she would gladly assist me in any other way she could. I was to write and tell her how much money I thought I needed, and she would do her best to find it.

Next weekend Daddy arranged for me to be driven to the Niokolo Koba guard camp, sixteen miles from Mount Asserik, to have a look at the area, meet the people staying there and make arrangements for bringing the chimps. It was a long, rough journey and we lost the way twice, so instead of arriving that same evening, we had to spend the night in the Land Rover somewhere in the middle of the park. We reached Niokolo camp, dusty and weary, late the following morning. For the most part the park vegetation was scrub savannah, but on the slopes of the hills and plateaus and in the small gullies that led into the river, the vegetation was thicker and resembled more closely the sort found in Abuko. I recognized several trees and plants that were food sources for the chimps. I tried to visit Mount Asserik but, as Mr. Dupuy had predicted, the rain had completely washed away the roads in some places.

We had decided that initially I should take only the three older chimps: Tina, now aged eight, Cheetah, seven, and Albert, also seven. They were the main troublemakers in the reserve, and would also be most likely to adapt to a truly wild existence. If all went well the others could follow when I was better equipped and when their seniors were well enough established to tutor them in the arts of their new life. The saddest separation would be that of Cheetah from William. For four years they had been close friends. I reflected a while and decided that though William was a good two years younger than Cheetah, I would take him with us. How much more comforting it would be for them to face the confusing transition period together.

BACK AT ABUKO, the day of departure arrived. With the aid of a large bunch of bananas I managed to persuade the four chimps who now lived outside in the reserve to enter the enclosure. The tranquilizing drug the veterinarian had provided was carefully measured out and given to Tina, Cheetah, Albert and William in some strong, sweet fruit juice. I sat on a bench while we waited for half an hour to pass and the drug to take effect. I noticed my hands were trembling slightly and that Daddy was perspiring freely despite the cool of the evening.

Fifteen, then twenty minutes ticked by and not one of the chimps looked less lively than it had before we'd treated them. In fact, if anything they seemed more active as they chased and tumbled with each other all round the enclosure. Then, forty-five minutes after we had administered the drug, William stopped playing and seemed to sway slightly as he approached the bench where we sat. He lay spread-eagled on his back and appeared to sleep. We went into the enclosure but he immediately sat up again, still far from being in a state where he would accept confinement in a crate. Two hours later William had recovered and Tina was asking to be let out of the enclosure. It was almost dark and only William had shown the slightest sign of being affected by the drugs.

The following afternoon we once again gave the chimps their drug, this time increasing the doses slightly. But although all the chimps became slightly uncoordinated and looked drowsy, I was sure that Tina and Albert were still sufficiently aware of their surroundings to make it impossible to confine them without a terrible struggle.

I had always been uneasy about the effect of tranquilizers on the chimps, and no one could persuade me to subject them to a third session. I let Tina and Albert out of the enclosure and cabled the park authorities that we were going to be late. The following day I sat on a log close to the traveling cage and tried to think of another solution. The fruit I'd put into the cage for the journey was still there, so I went over and took out a mango to suck on while I thought.

Tina was beside me before I noticed her. Pooh, Happy and Albert followed her out of the bush into the clearing and sat down around me, staring intently at my mouth while I ate. "There's a whole pile of fruit over there," I said aloud, and absentmindedly waved a hand in the direction of the cage. Tina suddenly began food-grunting and without any hesitation walked directly into the cage and gathered an armful of fruit. Albert, Happy and Pooh walked after her. Each took some fruit and returned to sit near me.

It took me an instant to realize what I'd seen. I had assumed that having spent days in filthy boxes before they arrived at Abuko, they would be exceptionally wary of entering a confined space again. Obviously I was wrong. During the years they had spent at Abuko perhaps those unpleasant memories had been erased; or else they had learned to trust us so completely that they no longer feared we would trick or hurt them. At any rate, I knew I had the solution to caging the chimps, and wished only that there was some way I could explain to Tina and the others that it was for their own sake that I appeared to be betraying their trust in me.

For the rest of the day the cage was kept filled with fruit. That evening I took all the chimps out together and led them to where I'd been sitting that morning. As I expected, all of them entered the cage confidently. Cheetah was the only one that hesitated slightly, shooting me a quick glance before walking into the cage to gather a share of the fruit.

We decided to try to trap the chimps on the following afternoon and to travel through the cool of the night. I would make the trip with John Casey, an officer at the reserve, and a driver, both of whom would return to The Gambia after a day or two. I planned to stay for five or six weeks.

That afternoon we took the chimps out for a short walk so that everything would appear as normal as possible. As we approached the baited cage, I lifted Ann and Flint into my arms while Abduli put Pooh on his shoulder and sat Happy on his hip. We had tied the fruit inside the cage so that the chimps could not immediately come out with it. The plan worked perfectly. Tina and Albert entered first; William followed. Cheetah again hesitated and looked around him, but then walked in to join the others. There was a sharp sound as John pulled the cord and the door of the cage slammed down, trapping all four chimps inside. Cheetah screamed harshly and banged on the bars, then ran to William and embraced him. Tina and Albert looked around in bewilderment. They came to the bars and, side by side, gazed out at us all.

While I waited for the Land Rover to arrive, I sat close to the wire and spoke to the chimps. William seemed perfectly at ease and sat eating, with Cheetah's arm wrapped around his shoulders. Tina and Albert huddled as close to me as they could get. Occasionally Tina would reach out, take my hand and guide it toward the padlocked bolts on the door. To make her understand that there was nothing I could do, I pretended to try to unlock the door but made it seem impossible. Then I hugged and reassured her through the bars, trying to tell her that though all this was a nuisance, there was nothing to worry about. I'd stay very near. How much she understood I had no way of knowing, but so long as I remained close enough for her to hold my hand, she stayed patient and calm. Cheetah, Albert and William modeled their behavior on Tina's, and so long as she remained unafraid, they were also able to relax.

There had been so much to occupy my mind through the day that I had had no time to become sentimental, but at seven thirty that evening, as I watched Abduli and the family saying good-bye to the chimps, the reality of what was happening struck me. A huge sense of grief and panic welled up, and I had to use all my self-control to avoid breaking down in front of everyone.

PART TWO

RELEASE

12. THE FIRST RELEASE

WE ARRIVED AT NIOKOLO CAMP the evening of the following day. As we could not set up our own camp on this occasion, we had to release the chimps near the main camp, a long way from where the wild chimps were usually seen. I chose a spot about three miles out from the guard camp, on the banks of the River Niokolo, so that there would be no problem for the chimps in finding water.

On the morning of the release we drove the Land Rover to the spot and the crate was opened. The chimps were ecstatic. They had hated every minute of the journey, though I did all in my power to ensure that they were comfortable, stopping regularly to give them sweet juice, milk and fruit and to clean out the rubbish. Now that they were suddenly out of their prison they rushed about hugging each other. Albert, in his excitement, mated with Tina before dashing over to hug and pat William.

I was frightened and anxious for them during those first few days. Tina was by far the most competent one among us, and so it was natural that she should take over the leadership. She blazed the trails, tried the new foods first, chose the nest sites and generally gave us all a little confidence, of which we were badly in need—perhaps I more than any of the others. However, on that first day even Tina seemed depressed and lethargic. Probably tired from the journey and confused by all that was taking place, the chimps appeared miserable. The vegetation was quite different and at first there seemed to be no food. The chimps certainly ate very little that day.

We moved on down to the river. William and Tina seemed afraid of so much water, but Albert and Cheetah took a drink, perched on overhanging branches. Tina then led off along the bank, and we followed. We came to some fallen rhun palm fruits that had split open from their long fall, but the chimps only smelled them before continuing. They looked bored and listless and kept asking us for food and water. Cheetah attempted a feeble little nest but quickly abandoned it. Tina tried it out a little later, adding to it, but only sat in it for three minutes before stretching out on a branch.

I sent the guard who'd accompanied us to the release site back to camp to bring the remaining mangoes. When he returned I gave the chimps about six mangoes each. William ate very little and seemed to be getting more listless as the day wore on. The others, though, were slightly refreshed by the meal, and when they had rested for a while Tina led on again. William, normally an energetic bundle of mischief, always in the lead on the Abuko walks, trailed farther and farther behind.

The fact that the chimps were eating made me feel a little brighter, but I kept wondering sadly if I was doing the right thing. Was I expecting too much to hope that they could make such a big break? What would become of them if they couldn't? The word "zoo" loomed up and made me shudder involuntarily. Slowly we climbed the slope to a flat-topped hill, which we christened Tina's Mountain. William sat close to me while the others explored. As the evening drew on, they wandered off along the edge of the plateau and disappeared. William stayed beside me, and eventually I picked him up and followed the chimps. I came across them sitting round a water hole from which they all drank except William. I filled our empty water bottle from the pool and offered him that; he drank greedily, almost emptying it. By this time the others had started out again. I stood and waited with William at the edge of the clearing, hoping he would follow them. Albert left the other two and came back to us. He wrapped an arm about William, who placed his hand on Albert's back, and two little white bottoms moved off slowly into the gloom. I should have been relieved—but I cried instead. We waited to hear the sounds of nest-making, then crept away.

Next morning, after a stormy night, I was up at four thirty. It was getting light and the air was fresh and clean. I scanned the slopes, but could see no sign of the chimps. Panicking a little, I walked up the slope, wondering if they had already risen and set off alone, thinking I'd deserted them. What of William, lagging behind the others? The higher I climbed, the more convinced I was that they had gone out into the unknown wilderness beyond. I cupped my hands to my mouth and shouted as loudly as I could.

Cheetah appeared first, from behind a boulder, followed by William, then Tina and Albert. William and Cheetah were very pleased to see me, and from their heavy lids and the sleep in their eyes I guessed I'd called them out of bed. All save William devoured the mangoes I had brought; Willie ate one or two but with little enthusiasm. None of them seemed any the worse for wear for their drenching during the night. One of the nests had disintegrated and a new one had been built, for there were five nests, all in the same tree.

By eight o'clock we were on the move, Tina leading. We explored the plateau thoroughly that second day, and I was relieved to see that there was a lot more food available than I had previously thought. William was really miserable all day and kept lagging behind. The adjustment would be difficult for all of them, but I began to feel that it was going to be almost impossible for William—he was too young and too unused to fending for himself. I decided that he should return to Abuko with John when he left the following day.

EACH DAY A PARK RANGER, or more often Réné, who was a laborer at Niokolo, would accompany me into the bush, as it was considered dangerous for me to wander about alone with the chimps. With William gone, things went better. The nests were now well made and the sites carefully chosen to give as wide a view of the surrounding country as possible. Whether this was instinct or thinking, I couldn't say, but even Cheetah, when choosing a nest site in the evening, would usually choose one high in a tree. The chimps were generally less fussy about day nests, and these were lower and sparser than night nests. Usually they would lie out on a bough in the heat of the day, constructing day nests only if the sky was dark and rain threatened, or to dry off in after a shower.

I hadn't noticed the chimps being so vocal at nesting time in Abuko. Here they seemed to communicate with one another with grunts when one of them decided that it was nesting time. They were sounds very similar to the "good food grunts" but never gave way to the food "chirrups and squeaks"—just the throaty grunts were used. Cheetah was especially noisy just before nesting and while constructing his nest. If one of the others answered him, his grunts would give way to ecstatic bouts of panting, and the others would join in for a second before getting on with the serious task of making a bed. Once they were all settled, and all odd leaves and shoots had been meticulously tucked into the platform, one of them would sometimes pant a "good night" and the others would answer. I got the impression that this was a means of checking that everyone was in position.

One hot morning there came a low warning hoot which brought all of us to immediate attention. Peering round my boulder, I saw that Cheetah was no longer stretched out but standing up, hair raised, staring hard at the other side of the clearing. Tina and Albert were soon beside him. I remained where I was behind my boulder. I couldn't see what was holding their interest at first, but then something the color of laterite moved, and out from the vegetation on the far side stepped a solitary male cob antelope. He had seen the chimps and was staring at them as he took hesitant steps into the open. Tina and Cheetah, hair sleek now, took little further notice once he was in the open, but Albert climbed onto a branch that jutted out into the clearing and shook it vigorously, giving a single hoot as he did so. The cob stood stock-still, staring, until even Albert lost interest and climbed down again. As he reached the ground the cob whistled once or twice, then daintily picked his way over the hard laterite table to the bush at the side and melted away.

It was heart-warming to watch the chimps meet neighbors: the slight initial hostility, then the realization, on both sides, that there was no danger, and the casual parting of ways. I really began to feel that this was where the chimps belonged. I was lucky to have shared their lives, but I had to accept the separation graciously and ease them into their new way of life.

By now we had been in Niokolo Koba about a week. The chimps were finding their own food and water and, although I still hadn't seen any signs of play among them, their initial depression seemed to be lifting a little and their awareness of the bush was increasing. Tina was exceptionally alert and wise, avoiding, if possible, open spaces and keeping to the trees. When startled by a noise she was up on a branch in a flash, from where she decided her next step. Fortunately the others heeded her every move and warning. If for some reason we had to cross an open grassy space, then every few minutes she would stand erect, or climb into small shrubs and trees and look all around her. Her actions made us all a little nervous, which in a way was a good thing, for after the safety of Abuko we tended to plod along rather complacently in her wake.

Unfortunately for the peace of our existence, a safari park in France had sent eight unwanted lion cubs to Niokolo Koba for release into the wild as soon as they were ready. The cubs, which were housed in an enormous enclosure next to the camp, now decided the time had come to release themselves. They were between twelve and eighteen months old and, though somewhat accustomed to man, would be unpredictable in their reactions if confronted, and still more so if they met a chimpanzee.

On several occasions one afternoon, while feeding, Tina and Albert stared down the slope, hair rising slowly then relaxing again, without having uttered the alarm hoot. At about four o'clock they both swung into a small tree, seeming nervous, and by means of overlapping branches crossed into a large fig tree and resumed feeding. As I approached, Tina and Albert moved off once again, and I presumed the figs could not have been ripe. Albert seated himself on a boulder in front of me. I looked up to find him staring at me. I held his gaze for a second before he glanced away, then watched him walk to the edge of the plateau and sit down.

I was getting up to see what they were eating when Tina came into view. Every hair on her body was erect, and as she stood upright she looked enormous. Her shoulders were hunched forward, her chin tucked in, and she held her arms slightly away from her body, looking for all the world like a character from an Italian Western about to draw as she swaggered forward. The sight of her made me stop in mid-crouch. I spun round so fast that instead of finding myself on my feet, I was still kneeling, staring, horrified, into sixteen curious kittenish eyes. On my hands and knees, rooted to the spot, I watched a magnificent Tina advance slowly. One by one the young lions turned and trotted across the open plateau into the vegetation on the other side. As they retreated Tina dropped to all fours, loped to the center of the clearing, sat down and hooted at the tufted tails. As though receiving a cue, Albert and Cheetah joined in from where they now stood, next to me. The spell suddenly broken, I scrambled to my feet.

The return to the wild produced some curious shifts in relationships. Once I saw Tina dig down through the peanuts and biscuits to extract the last mango, which had remained hidden beneath the other food. Holding it in one hand, she tentatively picked at the skin with the forefinger of the other. Albert watched her intently, then, much to my surprise, calmly stretched out his hand and took the mango. There had been no staring or hesitation and no hint of tension or aggression on either part. Yet Albert had always been very much the underdog at home. Even the youngsters, Flint especially, would take advantage of his submissive disposition, and though, toward the end, he had grown a little more confident, the older chimps—Tina, William and Cheetah—still bullied him. But since Albert had always been free at the reserve, fending for himself entirely during his first year, Niokolo Koba wasn't as strange to him as it was to Cheetah, who, except for the morning walks, had almost always lived in an enclosure and was far more familiar with and dependent on people. It looked as though Albert was beginning to make the most of his sudden advantage, whereas Cheetah, finding himself in a strange new world, had lost a lot of his former self-confidence.

During the five weeks I was with them I noticed that though Cheetah learned much from the others about self-preservation and became less dependent on me emotionally, he leaned more and more heavily on Tina. He often kept close to her by day and at night built his nest as near as he could to hers, whereas Albert would invariably choose a tree of his own nearby.

On another occasion Tina had taken more than her fair share of a bag of biscuits that I had brought. Albert took a look at the almost empty bag and the pile of biscuits in front of Tina, and walked over to her. Without any sign of hesitation, or begging permission, he stretched out his hand and appropriated some biscuits from Tina's pile. Tina snatched viciously at his hand. Grinning and squeaking, Albert turned to me. I made no movement, and I can't remember consciously changing my expression, but whatever he saw in my face gave him the courage to fly back at Tina. A violent squabble began, which lasted for only a few seconds before Tina, surprisingly, backed out. Sitting in the place where she had rolled and somersaulted during the fight, she watched a still grinning and squeaking Albert help himself to her biscuits. It was the first act of aggression I had seen among the chimps since we arrived, and it certainly looked as if meek, submissive Albert was gradually taking over as dominant chimp.

That evening found us peaceful and silent, near the edge of the ridge. The chimps were feeding in a grove of young kenno nearby while I sat propped against a boulder beneath the nesting trees, contentedly watching them feed. The ridge drops almost vertically for about the first ten feet, then slopes more gradually, so that when I heard a sudden deep grunting below me I couldn't see what had made the sound. The chimps climbed down silently from their trees and stared in the direction of the noise. At first I thought it might be warthogs rooting around. I crawled forward as quietly as I could, and had almost reached the ledge when a large baboon appeared a few yards to my left. We were each as startled as the other. If the shock of suddenly coming upon a crawling human hadn't given him cause enough to flee, the sight of Tina, hair raised, swaggering upright toward him, certainly did. The baboon ran frantically for the cover on the slopes. Tina barked aggressively as he sped away and gave chase as far as the edge of the ridge. There she stopped and sat down; I was soon by her side, and together we watched the troop of baboons disappear down the slope, barking intermittently in our direction.

More alarming encounters were to follow shortly. A day or two later I was squatting on the roadside, idly fingering some grass stems, when Cheetah's scream rent the silence. More screams followed, punctuated by aggressive barks and hoots. I found myself running as fast as I could toward the noise. Cheetah came into sight first, racing toward me, grinning. Behind him, Tina stood on a bank of earth, her hair raised. She was shaking a small shrub in one hand, stamping her feet and swaying from side to side. On the other side of the bank, barely five yards away, stood one of the young male lions from the safari park. His comrades were already making their way along the riverbank in the opposite direction. My presence beside her seemed to give Tina extra confidence. She made a swift lunge at the lion and only just missed a hearty pat on the back from his paw. He snarled as he struck out, obviously alarmed by Tina's sudden onslaught. This, and possibly the presence of a human so close by, made him turn and follow the other lions. Feeling they had won the day, all the chimps began screaming and barking.

If I expected them to be nervous after this episode, I was quite wrong. They seemed in unusually high spirits, and when a cob crossed the road ahead of us, instead of ignoring it as they usually did, Cheetah stood up, shook a shrub nearby and stamped his feet. Then, heading toward us, he continued to slap his feet down on the ground. When he reached us, he and Albert burst into the first signs of play I'd seen since our arrival.

We were soon to discover that there were wild lions in the area too. One evening I saw Cheetah hesitantly making his way up toward the plain above us, every hair erect. I began to follow, curious to find out what was alarming him. He had almost reached the edge of the vegetation when he suddenly flew round and came scrambling back toward me, grinning and squeaking in fear as he leaped into my arms. A female bushbuck came careering off the plateau. She seemed to be heading straight for us, but veered off to the left on sighting us. Close behind her came a male lion, bounding from boulder to boulder, his tail curved above him to balance him in his precarious pursuit. It was all over in a few seconds, with only the crashing of vegetation on my left to convince me I had not been dreaming, but even now, if I close my eyes, I can see that lion leaping the rocks on the slope above, his light-gold mane bouncing on his neck. I don't think he saw us, he was so intent on his prey and the rock-strewn path.

As the noise of the chase faded into the distance, Tina and Albert climbed down from their tree and Cheetah left me to join the other chimps. Nervously, they made their way to the edge of the vegetation. They certainly showed no signs of wanting to follow the wild lion we had just seen—in fact the event seemed to have unnerved them, as they peered out of the vegetation onto the plateau.

The chimps were now beginning to hunt again. I was waiting with Cheetah while he finished drinking when I heard one of the chimps scream, not with terror but with rage, followed by a volley of aggressive barks. I hurried to the scene. Albert was sitting on the ground, holding a young green vervet who was approximately 2–3 months old. The sight was a gruesome one. Albert held the vervet's head and shoulder with both hands while he tore at the face with his teeth, occasionally spitting out small pieces of bone. There was none of the hysterical excitement I had witnessed at other kills. Albert ate the whole face and the brain, which, from his chirrups of delight, I think he really enjoyed. He then abandoned the monkey—almost unmarked, apart from its empty skull—and went over to where Tina and Cheetah still sat. The other two were not visibly agitated as he approached, and soon all three were feeding on a fig tree, the monkey apparently forgotten.

Réné, who was accompanying me, told me what had happened. After they had left Cheetah and me at the water hole, Tina and Albert had surprised a troop of green vervet monkeys feeding in one of the many fig trees which grew on the grassy strip of land between the road and the denser slopes of Point Plateau. The youngster they caught had tried to make his escape by running across a large open patch of grassland. It was here that the chimps caught him. Apparently Tina had held his head and Albert his hindquarters. The young monkey's mother, on hearing her offspring's shrieks, had left the tree and made a desperate run at the chimps, but her courage gave out before she reached them—and she raced back to the safety of the trees, where she sat until I arrived. It was Tina who had killed the monkey, biting it several times in the head and side, and then walking off and allowing Albert to have it. There had been no squabbling between the chimps over the possession of the spoils.

Albert was very playful all that evening and kept inviting Cheetah to a game, but got only a halfhearted response. All the chimps climbed into a rambling fig tree and either fed or rested. Albert began to groom Tina, whereupon Cheetah immediately climbed down and headed purposefully toward me. Tina followed him and a few minutes later Albert followed Tina, and they all sat round me. Albert edged over to Tina and once more started to groom her, at which point she crouched and Albert mated her. Cheetah sat up then, and with his hand raised above his head, waited for Tina. She didn't respond immediately, so he began to whimper as added persuasion. Tina left Albert and went to Cheetah, who mated her, then possessively groomed her, glancing frequently at Albert.

It began to rain hard. The chimps huddled miserably beneath a tree, but as the rain subsided they were soon playing energetic games—to keep warm, I suppose. Chasing each other, wrestling while perched precariously on boughs, they just generally had fun. Albert kept hanging upside down, holding a branch with both feet and panting to himself, at which point Cheetah would do his best to pull him down out of the tree.

After about an hour they seemed to have worn each other out, and both became a little more subdued. Cheetah built a small nest fairly high in the tree. As soon as he had settled back on it, though, Albert began to bounce the bottom of the nest with his knuckles, pulling away at the foundation leaves and branches. Good-humoredly, Cheetah leaned out, panting, and they began to grapple and wrestle again till the nest had a huge hole in the bottom of it. Cheetah stuck his head through it, and for the next few minutes their game consisted of sliding through the nest headfirst, which they both found hysterically funny. Tina sat in her fig tree and watched, huddled up and cold. I was glad when Cheetah chased Albert into her tree and she joined in the circus. Réné went over to watch them more closely; as he stood beneath the tree, Cheetah rushed to the end of the branch and shook it, showering him with water. He must have thoroughly enjoyed Réné's reaction, for he (and later Albert) kept shaking rain-soaked branches over our heads whenever the opportunity arose, panting as they rushed off again.

THE COMPETITION for Tina's attention was growing fiercer. One evening Albert came down from his tree, approached Tina and was about to mate with her when Cheetah rushed up to prevent him. They had a very aggressive tussle, filling the air with screams and dung. It was Cheetah who backed out, even though he was slightly bigger and heavier than Albert. He came loping toward me, grinning and squeaking, seeking reassurance. He put his arm around my waist and buried a grinning face in my shirt. I comforted him cautiously; I didn't want him to draw courage from my sympathy and begin the fight afresh. He remained seated close to me, looking very glum as Albert mated Tina and then began grooming her. I groomed Cheetah to help him calm down.

A few minutes later Tina and Albert were feeding from a shrub, seated side by side on the ground. Cheetah approached them and, very cautiously, began to groom Tina's shoulder. Albert made no objection, neither did Tina, so Cheetah continued. Soon afterward Albert got up and walked away. Seconds later Tina followed him. As she rose Cheetah sniffed her pink bottom, but made no attempt to mount her. They all climbed into a fig tree and fed peacefully. I noticed a nasty bite on Albert's leg, which was bleeding. It had probably been inflicted by Cheetah in the tussle earlier on.

Next day the competition was renewed. Albert climbed over to Tina and mated her. On seeing this Cheetah came careering down, squeaking at Albert, who took no notice. Cheetah stopped a few yards from the offending couple and sat down, whereupon Albert climbed on Tina and mated her for the second time within the space of a minute. It was if he wanted to prove something to Cheetah, who stared hard but made no aggressive sound or action. When Albert had quite finished, Cheetah reached out a tentative hand to Tina, who immediately moved out of the way, up another branch. Albert followed her and then began to groom her. Looking very depressed, Cheetah remained where he was, pulling at his penis as though it were to blame.

Tina's pink swelling gradually subsided and the three chimps became good friends once more. At the beginning of the fifth week I spent with the chimps we came upon an enormous grove of figs, and for the rest of that week they did little other than stuff themselves. They were also extremely good at finding water holes unaided.

I stayed with the chimps five weeks. By the time it was necessary for me to return to The Gambia, I felt confident that Tina, Cheetah and Albert had made the adjustment from the semicaptive state of Abuko to true independence in Niokolo Koba.

Even so it was a desperately sad moment for me when I had to leave them as they fed heartily in a fig tree. I wondered when, if ever, I would see these three very good friends of mine again.

SHORTLY AFTER MY RETURN to The Gambia I received a letter from the park authorities telling me that ten days after my departure Cheetah had returned to camp alone, and remained for four days. He had evidently left Tina and Albert or had been driven away by them, and was going it alone. The park warden had driven him toward Mount Asserik and left him there.

Despite my responsibilities at Abuko, I returned to Niokolo Koba at once and spent a week searching for Cheetah and the others, but with no success.

During the next year I visited Niokolo Koba whenever I got the opportunity, but I was never able to stay long enough to make a really thorough search and I found no trace of my chimps. All that was left was the hope that Cheetah would someday meet wild chimps and tag along with them until he was accepted. As for Tina and Albert—at least they had each other.

13. DEATH COMES TO ABUKO

A WEEK AFTER MY RETURN, Daddy left The Gambia for a stay in England. I was to manage Abuko in his absence. It would soon be time to open the reserve to visitors, and there were many half-finished projects to complete. The Abuko staff had just begun to dig an extra pool for the sitatunga antelopes in a dry section of the reserve. The rainy-season growth had to be cut back from the paths, a new photographic blind was to be finished near the lake, and in a particularly swampy area of the reserve a small wooden bridge needed to be constructed. There was also the varied collection of animal orphans: three antelope fawns, two more young hyenas, three warthogs, two baby monkeys, four civet kittens and a milky owl chick, all with their own special milk formulae and diets.

Luckily for me, the burden was to be shared. Nigel Orbell, who had worked with me during my stay at Woburn, was coming out to The Gambia to replace John Casey. As soon as possible after his arrival I drove him out to the Abuko Reserve.

We walked up the path so that he would get an impression of the vegetation before entering the Orphanage. The chimps began to hoot and scream excitedly on seeing us, and I promptly took Nigel over to meet them. Happy and Pooh lost no time in clambering up my legs to their usual positions, from where they extended the backs of their wrists to Nigel in friendly greeting. Ann sat slightly aloof, but Flint left her side to hurry over to us, his coat bristling with excitement. Hooting, he gripped Nigel's leg. As Nigel stooped to return the greeting, William leaped off a high crossbar and landed heavily on the newcomer's back. Nigel lost his balance and had his face pushed firmly into the soil. William then flew up the apparatus again, panting his delight at the result of his jest. Nigel picked himself up from the ground, trying desperately to remain composed, but William quickly snatched his glasses and fled. I gave chase, but on his next circuit around his floundering victim, William efficiently whipped the wallet from the back pocket of Nigel's shorts, ecstatic over his easy success. Nigel was virtually helpless without his glasses, and though furious and humiliated, he could do little more than blink his eyes and spit the dirt out of his mouth.

During the following weeks Nigel spent much time with the chimps. Soon he was able to take them on walks into the reserve on those occasions when I had to be elsewhere. Having learned what he could expect from William, he never gave him the chance to repeat the performance of their first meeting. He was strict but fair, and William quickly found that he could take few liberties with him.

One day Nigel came back from the reserve to report that Happy and Flint had developed coughs and runny noses and seemed very lethargic. That same evening they refused to eat anything and had high temperatures. I took them home with me and put them on my bed, giving them warm milk and half an aspirin each.

During the night their condition deteriorated rapidly, and I realized that they were suffering from something far more serious than the average cold. They lay beside one another, still and listless, their breathing fast and shallow, their temperatures remaining high. When I roused them for warm milk with honey, Happy's long eyelashes flickered open briefly and his doleful brown eyes were bright with fever. On seeing the cup, he turned his head away weakly and closed his eyes in silent rejection. Flint could be persuaded to drink only a few mouthfuls.

Late as it was, I telephoned a friend of ours, Dr. Bradford, a pediatrician. To my relief he said he would drive to Yundum straight away. Less than an hour later I heard a car stop outside the garden gate. As we walked into the house Dr. Bradford listened carefully to my description of the symptoms Happy and Flint were showing. At the end of his examination he diagnosed virus pneumonia. He told me to keep them under close observation and gave me antibiotic drugs and some tablets that would help their breathing.

At daybreak Flint looked as if he had improved; he would take drinks, so getting him to swallow the bitter drugs was not too difficult. He still had a bad cough, though, and scarcely any appetite for solid foods. Happy was definitely no better; he lay limp and on the verge of unconsciousness, refusing to drink even the smallest amount of milk.

When the doctor returned, he advised that we use injections rather than oral treatment. Happy barely flinched as he was injected, but not so Flint. Weak as he was, he fought like a demon and, considering his small size, showed incredible strength. He bit and screamed as I struggled to hold him still enough for the needle. His resistance was obviously not helping his condition, so in the end we had to continue treating him orally. The fact that I had been obliged to use force with Flint had badly frightened him, undermining much of his former confidence in me. It was disconcerting when, later in the day, I held out my arms to him and, instead of climbing me as usual for comfort, he backed away, grinning and whimpering his distrust. I had to spend hours coaxing him before he would approach me voluntarily, and even then the relationship was fragile and could easily be shattered if I was not careful. Thereafter, persuading Flint to drink his medicine, heavily disguised though it was, became more and more difficult. Somehow, though, we managed to give him the prescribed dosage.

Despite the injections, Happy was no better. He looked pathetic lying on my bed, his dark head propped up on a great white pillow, his eyes reflecting the extraordinary effort required to breathe. Every few hours during the night I would check on them both, and each time I would approach their bed with mounting dread until I had leaned over them and satisfied myself that they were asleep and as comfortable as I could make them. In the early hours of the following morning I hurried over to the chimps. Flint stirred a little at the sudden brightness and then went back to sleep. Happy was unconscious. His silky lashes half masked his upturned eyes, and from his sagging mouth came the rasping sound of his labored breathing. There was little I could do other than change his position in the hope it would make breathing easier. In desperation I took one of his limp hands and tried to will all the energy and health I possessed into his exhausted, heaving body.

When the doctor arrived next morning, I was still holding Happy's hand and Flint was asleep on my lap. I stood up stiffly, carrying Flint, who began to struggle at the sight of Dr. Bradford. The doctor gave Happy two more injections and showed the position in which it was best for him to lie. He advised me to keep some cotton wool handy to wipe away any excess fluid that might appear from Happy's nose or mouth. Dr. Bradford slowly extended his hand to Flint in an attempt to be friendly, but Flint in sudden panic wrenched free of my arms and ran screeching to the other end of the room. It was encouraging that he was able to object so strongly, and although an examination was obviously impossible, he confirmed that he was feeling better by eating some banana and papaw for lunch.

The bedroom was a hushed, depressing place. I followed the doctor's instructions, and when I could do no more, simply sat beside Happy and waited. By early afternoon I thought I could detect a slight improvement in his breathing. A little later he began to stir. Cautiously, I allowed myself to hope that he was getting better. Happy's eyelids began to twitch and he closed and opened his parched mouth. I took his hand again and called his name softly. Hesitantly, his eyes opened and he gazed about him wearily. I continued whispering to him in case he should be afraid. When his attention drifted to my face, I saw his lips form an "ooh" of recognition.

When Dr. Bradford called again that evening, Happy was sleeping. I woke him for his injection, and he held his arms out weakly to be lifted. For the first time he seemed to feel the injection, for he gripped my hand tightly and pouted. He accepted a drink and then fell asleep again. Flint was well enough to accompany us onto the verandah, where Heather had set out some sandwiches and drinks. Flint had a cup of lime juice and honey and shared in the sandwiches, keeping a suspicious eye on Dr. Bradford.

Two mornings later I woke to find Happy sitting up in bed chewing on the remains of Flint's supper, most of which was smeared all over the blanket. He greeted me with a few nasal grunts. Flint was missing. I checked the open wardrobe without success, then noticed that the bedroom door, usually kept closed, was ajar. In the kitchen, the door of the refrigerator was open and most of its contents splattered liberally over the floor. Flint was sitting on the kitchen table, chirping and grunting to himself over the last of a large jam tart. I salvaged all that I could, and carried Flint and his piece of tart back to the bedroom. By the end of the week I thought him well enough to return to Ann in Abuko.

When Happy's series of injections finished, he was considerably more active and readily consumed all the fruit that Dr. Bradford had left him. It was lonely and frustrating for him to be confined to the bedroom, for every time he found something to play with, such as an alarm clock, picture frame, clothes, or light switches, I was forced to stop him. Soon he, too, returned to the enclosure. I was thankful that the anxiety-filled days were over.

But less than a week after Happy's return, I noticed that Ann had become listless and had lost some of her appetite. She had a runny nose, and Flint, who was, as always, close to her, looked very subdued as well. I began to treat them with the oral antibiotic that Flint had been receiving previously, but by evening they were coughing and seemed worse. They were still active, though, and eating small amounts of fruit and medicated juices. I began to wonder whether I should take them home, but decided to leave them for another night in the hope that it would not be necessary to remove them to the intensive-care unit of the bedroom.

I arrived at the reserve earlier than usual the following morning and went straight to the store to make up a cup of fruit juice for Ann in which to disguise her medicine. William, Pooh and Happy met me at the enclosure gate, but Ann was lying on her stomach, her face buried in her folded arms. I thought it strange that Flint was not near her and presumed he must still be in the hammock. I walked up to Ann, telling her to drink up her juice like a good girl. She didn't stir. I squatted down and talked to her, but she still gave no response. Her face was hidden, and believing she was still asleep, I stroked her shoulder to wake her. As soon as my hand touched her body I realized with a sickening jolt that something was horribly wrong. She was limp and heavy as I gently lifted her up. I clutched her tightly to me for several minutes before I could begin to understand the truth. Ann was dead.

I wiped her nose and mouth with my shirt, and closed her eyes. Then, almost unable to see where I was going, I carried her toward the enclosure door. Abduli met me there and took my hand.

"Flint," he said sadly, "Flint very sick too."

I raised my streaming face to look at him, and asked chokingly, "Abduli, you knew?"

"Ann die before I come today. I find no way to tell you, Stella. I take you Ann now, go find out Flint."

Dazed, I walked back to the enclosure and climbed the crude ladder to the sleeping hut. Flint was in his hammock, fighting for breath. I carried him down, and he hung as limply in my arms as Ann had. Happy was waiting at the bottom of the ladder, gazing upward. He reached out as I passed, and with slow, spidery fingers briefly touched Flint's foot. Pooh sat quietly beneath the feeding table, but William bounded over, pulling playfully at Flint's dangling hand. He looked puzzled when I pushed past, ignoring him.

As I approached the rest house, the Land Rover was pulling up outside the orphanage. Nigel got out cheerfully, slamming the door behind him. He strode into the orphanage and then stopped, seemingly paralyzed, a few yards in front of me. His expression changed as he absorbed the scene before him.

A moment later Flint began to tremble. I moved away from Nigel and sat down on the rest-house bench, cradling Flint's shuddering body. It was too late to do anything other than give him all the warmth and comfort I could during those painful last moments of his tragically short life.

At midday all those working at Abuko pushed their way through a curtain of leaves and twisted vines to a small clearing which was kept shaded and secret by an ancient spreading mandico tree. We stood solemnly in a semicircle as Ann and Flint were laid side by side in their grave and buried.

The enclosure looked as painfully empty as I felt that evening. As I gave William, Pooh and Happy their supper, I wept bitterly. Ann and Flint could never take part in the plans and hopes I had of true freedom for all of the chimps in Niokolo Koba.

14. A TIME OF PREPARATION

DURING THE DEPRESSING WEEKS that followed Ann and Flint's death, I spent most of my time at the reserve with William, Pooh and Happy. It was clear that they missed the others a great deal—William especially. Pooh and Happy had always been together and, since they were about the same age, they had an ideal playmate in each other.

William and Ann had spent almost five years together, and since Cheetah had left the bond between them had become more apparent. William, though noticeably subdued, tried to play with the little ones, but at that time the difference in age and size was too great for him to bridge successfully. He was often tough in play, which would make the little ones run to me whimpering. Also neither Pooh, now 3½ years old, nor Happy, 2½ to 3, was at an age when grooming interested them, and William missed being groomed. He would often invite me to groom him by placing himself squarely on the ground in front of me, but though I tried, I don't think I did as satisfactory a job as Ann or Tina would have. William still chased the colobus with enthusiasm, but though Pooh and Happy gave him all the vocal encouragement they could, neither of them was much use at helping him.

Often during walks I would watch William leaping around in one of these hunts. The first few times he missed a monkey he would whaa aggressively and try again. Gradually his reaction to the colobus's escapes became one of frustration; he would sit for a moment, whimper and shake his hands. After about an hour with no luck, it was not uncommon for William to throw himself down a tree and have such a temper tantrum that he would turn somersaults and scream until he choked. When he had calmed down he would ignore the colobus for the rest of the walk, but the following day would go through the whole process again.

I thought he would learn that to hunt alone was impossible and thus spare himself much disappointment, but William proved admirably persistent. Day after day he would hunt, until finally he was successful. He managed to seize a mother with a young infant. The mother escaped, but William killed the infant. I watched closely, expecting that William would at last imitate Tina and Albert's example and eat the monkey, but although all three of the chimps showed great excitement and frequently bit and pulled at the skin of their prey, none of them actually ate any. Instead, once the excitement of the hunt was over, they used it as a toy.

Following his first solo success, William hunted more avidly than ever, and before the end of the year had caught and killed four more colobus infants. Perhaps because he lacked playmates, he began to interact more frequently with the other animals in the orphanage. He would sometimes enter the enclosure where we had had to confine Charlie Oribi. William would dash from one end of the pen to the other with Charlie in hot pursuit, then leap up onto the wire to safety, leaving Charlie to butt out his rage on a shrub or any other suitable object.

William also discovered that Bookie and Buster, the hyenas, were exceptionally energetic playmates. I was anxious the first time I discovered William in the hyena pen, for the hyenas were adult now, no longer the boisterous cubs he had chased round the lawn in the past. Abduli and I tried to entice him out, but because he was having such fun, and perhaps because he knew we were worried, he was typically contrary and we could not get near him.

Fortunately the hyenas showed no inclination to be aggressive, and indeed were enjoying the game as much as William. Gradually both hyenas and William became more confident, till the hyenas were rolling on their backs while William tried to pull them along by their tails, or William was allowing them to catch him and playfully drag him about by an arm or a foot. Daddy and I had frequently played with the hyenas, and they seemed to be reacting to William in much the same way. Indeed, I'm sure they got more pleasure from cavorting and tumbling with William, for he was so

Stella with Bookie (top) and Buster (bottom)

Stella and Buster at play

quick and lively. He soon established himself as the dominant member of the play group, and the hyenas showed him much respect.

More and more my mind was turning to the problems of taking Pooh, William and Happy to join the others. Their rehabilitation would be a completely different matter from that of the earlier group. These three had all been captured at a very early age and, apart from life at Abuko, had no experience of the wild or of being independent. I knew I would have to spend a great deal more than five weeks on the second rehabilitation project.

It could take years. To be able to carry out my plans I would have to find funds from somewhere. I needed a camp of my own, a vehicle, equipment, to establish the rehabilitation center. I wanted the time to make a thorough search for Tina, Albert and Cheetah. I was certain that in time I could re-establish contact with them. By then they would be experienced in the ways of the bush, ideal tutors for Pooh, Happy and William, and the original group, though tragically two short, would be together again. Somehow I had to earn some money.

In reply to the report on the first three chimps I had sent her, Jane Goodall wrote inviting me to visit Gombe Research Centre in Tanzania. She had mailed to her London publishers, Collins, a long account that I had done for her of the weeks at Nikolo, and there was a suggestion that I might like to attempt a book.

I spent two and a half of the most exciting months of my life at Gombe, where I gained valuable insight into the lives of wild chimpanzees. Though much of their behavior was familiar to me, I learned more about the significance of each gesture and facial expression.

It was the mothers and their infants that fascinated me most. I had often constructed in my mind a picture of a mother chimp in action, and in dealing with the orphans at home I had tried to behave as much as possible in the same way. Now I was able to do something that I had desperately wanted to ever since I first picked up and tried to comfort William on his arrival at the office, years before. Feeling almost invisible, I observed a group of mother chimps with their offspring. The patience, protection and, above all, the comfort that the mother provided for her baby was everything I had imagined. To see the baby's needs correspond so perfectly to what its mother could give, the harmony that was formed by all the elements of their behavior, made me realize yet again, but with a more profound understanding, just how deprived were the chimps at Abuko. We had been able to give our orphans perhaps the best

Stella and Jane Goodall at Gombe

substitute lives that humans could provide, but it was still a poor exchange for their natural mothers in their natural homes.

One evening after we had finished supper in the communal dining room, I talked with Jane and her husband, Hugo van Lawick, about the plans I had for the remaining chimps at Abuko. I explained that I wanted to spend time in Niokolo Koba searching for Tina, Cheetah and Albert, and that when I had found them I planned to set up a campsite from which I could release William, Pooh and Happy. We discussed all the practical details and difficulties. I would have to have a Land Rover, or some other sturdy cross-country vehicle, and they told me from their own experience all that I would need to establish a very primitive camp. Later that evening Hugo helped me to draw up a budget for the project. Despite strict economizing, when all the figures for the first year were added up, it seemed an enormous sum of money. But Jane and Hugo were optimistic that I would find the funds I needed.

My stay was drawing to a close when I received a letter from my father which shattered the spell that Gombe had cast over me. While out on a walk, Happy had touched one of the electricity cables that ran across the end of the reserve and had been badly burned. Miraculously, he was still alive when Abduli picked him up, though his right hand was in poor condition. Daddy said that I was not to rush home: the accident had occurred almost a week ago, and with intensive nursing, Happy was showing great signs of improvement. Everyone believed his life was out of danger and that he was beginning to recover rapidly. I felt a long way from home, and wanted to be near Happy to nurse and reassure him.

A few days later I got news that Cheetah had returned to Niokolo Koba camp, alone and suffering from serious diarrhea. He had been fed and treated with the appropriate human medicines, and after four days appeared healthy once more. On the morning of the fifth day, Cheetah left. He set out along the banks of the River Niokolo, and when last seen was already a good two miles from camp.

Cheetah had appeared in Niokolo Koba at the end of July—exactly one year after he had been deposited near Mount Asserik. During that year no one had seen him. It seemed incredible that after a year of living independently in the bush, Cheetah should reappear at a camp full of people when he was feeling sick, and leave again so readily as soon as he recovered.

I was immensely encouraged to know that he had successfully survived his first year in the bush. He had undergone all the seasonal changes of diet and water sources and must now be well equipped with experience for the years to come. Only one thing worried me: he was still alone. I was sure, though, that it would not be long before I would be back in Niokolo, properly equipped to relocate him, to offer him the company of William, Pooh and Happy and perhaps, later on, to introduce other young chimps that could be saved from captivity. The thought made me more determined than ever to find the funds to make this possible.

When I left Gombe I spent a couple of weeks in London, trying to find financial backing. Hugo van Lawick was in town at the time and helped me a great deal, but there were few organizations interested in sponsoring my project, and I began to despair. Then suddenly everything seemed to happen at once. My future publishers, Collins, had been sufficiently impressed by what I was doing to risk putting up an advance that would finance the first year of my project. Hugo decided that he would like to make a film of it, and the advance on film rights was enough to cover the cost of a second year. The excitement I felt was indescribable. I could go ahead with my project, and one of the best animal photographers in the world was going to film it!

I shall never forget the delight of buying my first car, an ex-Irish Police Land Rover. Until then I had not realized what affection an inanimate piece of machinery could arouse. I fondly christened it Felicity.

While in London, I visited Dr. Michael Brambell, curator of mammals at the London Zoo. Dr. Brambell showed me two juvenile chimps, a male and a female. They were called Yula and Cameron and had both been born in London. They recognized Dr. Brambell and pushed their long pale hands through the bars to touch his face. Cameron, age 4, had been abandoned at birth, and Mrs. Brambell had raised him in the house till he was old enough to be put back in the zoo and cared for by the keepers. It had also been necessary to take Yula away from her natural mother, and from the age of about a year the two little chimps had lived together in the same cage. Cameron was the more confident and playful; Yula, age 4, was smaller, extremely beautiful and quiet. Her movements were slower and more deliberate than Cameron's, and her quiet, shy, yet determined manner vividly recalled Ann. Cameron kept tapping and grabbing at me playfully through the bars, but Yula fixed me with a resolute stare and began to groom my face. Any small blemish or irregularity of my skin was firmly picked at with the long nail of her forefinger.

To my mingled delight and alarm Dr. Brambell explained that Yula and Cameron could not be integrated into the group of chimps at the zoo. Did I think, he asked, that it would be possible to rehabilitate zoo-born chimps in the wild? The idea excited me immensely. Was it possible? Certainly Yula and Cameron would be ideal subjects for such an experiment. They were of West African origin and approximately the right age.

It was decided that as soon as I had found the funds to establish the camp and was able to see how my younger chimps were coping with rehabilitation, I would contact Dr. Brambell again. Further plans could be made for Yula and Cameron at that time.

Then came some shattering news. One afternoon my father telephoned me from The Gambia. He coughed slightly, then said that he had some bad news for me. I tightened my grip on the phone and waited. It was Happy. He had seemed to be recovering well from his accident, his hand had healed and he had started playing normally with the others. Recently, however, he had begun to lose his appetite. Nigel had tried every possible food, but even if Happy ate, he invariably vomited. The vet could not tell what was wrong with him. Then, during the last week, Daddy and Nigel noticed that he was peering closely at the surface of the table during the evening feed and that he was having difficulty coordinating his hands to pick up the small amount of food he ate. He had become very listless and dependent on others. The moment Nigel put him down and moved a short distance from him, he would look around wildly and begin to whimper in terrible distress. He no longer played and frequently sat staring into an unfocused distance.

After a short silence, I heard Daddy's voice continue softly: "Stella, try to get back as soon as possible. Happy is almost totally blind. He looks pathetically lost, and if he doesn't begin eating soon, I don't think he's long for this world. Come home quickly, the little fellow needs you."

For about ten minutes after I had replaced the receiver I was oblivious of everything except total, tearing misery. Happy's round baby face, framed in thick dark hair, appeared in my mind. I saw him staring up at us seriously with enormous eyes from the depths of a large, round basket; peeping at us over the edge of Tina's nest, while he clung to her long coat possessively, tumbling and wrestling with Pooh and Flint, panting hysterically with hoarse laughter; sick on my bed, holding out his arms to me and whimpering to be held. Then I saw him, small and alone, bewildered by the unfathomable darkness around him.

I flew back on the first possible plane. I had convinced myself that once I was home there would be something I could do for Happy. I would fly him to England for an operation by the best vet, if by these means his eyesight could be restored and his illness diagnosed and cured.

A little over an hour after arriving at the airport, I was at the reserve. Nigel had taken Pooh and William out for a walk and the enclosure seemed empty. Abduli told me that Happy spent most of his day lying in a bunk in the sleeping hut. I went into the enclosure and walked quietly to the ladder leading up to the entrance of their stilted dormitory. Happy had heard someone arriving and was sitting on the first rung. He was a shocking sight, small, thin and thoroughly dejected. I spoke his name and asked if he remembered me. His response was immediate and deeply touching. His pinched face broke into a grin, and almost screaming, he half slid, half climbed down the ladder toward me. I noticed his left hand was little more than a rigid claw which he kept pressed to his chest.

I climbed a few rungs to meet him, and we embraced each other emotionally. I sat and spoke to him for a while, stroking and caressing him. His eyes were as large and brown as ever; only at certain angles could I see the milky whiteness that was obliterating his vision. His gaze was distant and vacant. Apparently Happy's world was already one of vague shadows, for he made his way about by feel and memory.

I took him out of the enclosure to the rest house and offered him the fruit I had brought him from England. He seemed to like the grapes most of all, and Abduli and I were hopeful that at last we had found something he would take. Our satisfaction was short-lived, for about half an hour later he vomited up all that he had so slowly eaten.

While sitting in the rest house with us he seemed to be constantly listening for something and facing the entrance. Suddenly he grinned and gave his throaty whimper. A few seconds later I heard the orphanage gate open. Happy's reaction was one of excitement: he held up his arm and moved his head round anxiously, begging to be lifted. I picked him up and carried him out to meet the others. Pooh and William greeted me warmly—Pooh, like Happy earlier, almost screaming his excitement. Happy panted and held out his hand. I guided it till he was able to touch Pooh and William.

It was essential that we take blood samples from Happy as quickly as possible and send them to England for analysis. I had expected that I would be able to persuade Happy to remain still while the blood samples were being drawn, but he was not at all cooperative. Unable to see what was happening around him and hearing a stranger move nearby, he became suspicious and frightened. Finally the doctor suggested that he be tranquilized. Nigel and Abduli held Happy for the injection, and I picked him up as soon as it was over. He stopped screaming and clung to me while I walked round patting and soothing him. Minutes later he rested his head on my shoulder and slept. The blood samples were taken and rushed to the airport.

An hour after his injection Happy was still soundly anesthetized. I drove him home carefully, and for the second time in his short life he lay unconscious on my bed. Long after the plane had left Gambia with the blood samples, Happy was showing no signs of waking. Worried, I rang the vet and asked his advice. I was told that because Happy was weak, the drug might take longer than usual to wear off. Darkness fell and there was still no improvement. By 10 p.m. I was sufficiently alarmed to bring my father to Happy's bedside. Daddy placed his large, square hand over Happy's chest, then slowly moved it up to stroke his head. With startling suddenness Happy sat bolt upright, opened his eyes and coughed once, then fell back on the pillow. I leaned over quickly and began speaking to him, but Happy made not the slightest gesture in answer. Again I saw my father's hand carefully cover the small, still chest. I sensed rather than saw him become motionless in his concentration to listen and feel. Even before he relaxed again, I knew Happy was dead.

The following morning a cable arrived from England. Happy had had diabetes. Had he lived, he would always have been blind and would have needed insulin injections every day. He would never have been able to participate in the project, but would have had to remain all his life in an enclosure. The memory of his wide-eyed innocent face was painful, but when I read the telegram I no longer wished that Happy could have survived.

15. RETURN TO MOUNT ASSERIK

PREPARATIONS NOW PROCEEDED RAPIDLY. I planned to leave Gambia for Senegal at the beginning of January 1974, search for Tina, Cheetah and Albert till March, and then go back to Abuko to collect William and Pooh. Momado, our gardener, was to accompany me. I had decided to spend the first week near the village of Wassadon, on the banks of the River Nierriko. Wassadon was outside the park and several miles from where the chimps had been released, but there had been recent reports that chimps which seemed unusually accustomed to humans had been sighted in the area.

The first day there I found five fairly fresh nests, all about two days old, in the trees on either side of the bridge across the river. It was frustrating to think that I'd missed the chimps who had built them by such a short time.

The following day I walked up the Nierriko. I found a few very old nests but nothing else encouraging. During the day I'd met several people collecting firewood or felling palm trees for timber. No one spoke English, but between Momado and myself we managed to inquire whether they had ever seen chimps. One answered that there were hundreds of them, but I found that he had misunderstood us and was referring to the baboons. Another said he had seen chimps a few times but they only passed through the area. He thought they lived nearby on the River Gambia.

For the next three days I walked around exploring likely areas. Each day I found one or two nests, but they were nearly always very old. Along the River Gambia there were fresher traces of chimps, but even these places were plainly long deserted.

Frustrated, we packed up the camp and moved on to Niokolo Koba. The same afternoon I walked along the River Niokolo and up to Tina's mountain. It was still possible to see the remains of the nests that Tina, Cheetah and Albert had built almost two years before, and these tangible reminders brought on a mood of nostalgia and made me more anxious than ever to know how and where they were. I toured the plateau and went a little farther up the river, but saw nothing to indicate that chimps had been there recently.

For the rest of January I searched in the vicinity of Niokolo camp, but with no success. However, from the nest counts I discovered that wild chimps visited areas a good deal closer to Niokolo camp than had previously been thought.

During February I planned to explore the region of Mount Asserik. The land around the mountain is more undulating and picturesque and slightly greener than most other parts of the park. However, from the road the country still seemed parched, and apart from the thicker vegetation around the dry stream beds that crossed the road, I could not, at first, see that the area was any richer in fruit trees than that around Niokolo.

A mile or two before we began the climb to the summit, I noticed, on the other side of a vast, barren plateau, the tops of some green trees, two or three of which looked like "fromagers," or kapok, which usually grow near water. We drove across the plateau toward the line of green and discovered a deep, rocky gully which obviously held water in the rainy season but was already dry. The transformation in the vegetation was a thrilling surprise. Huge kapok and other sturdy, leafy trees formed a shady canopy above the gully, and as we descended into it, a drop in temperature was noticeable. We scrambled over the boulders and crevices of the gully till we found a well-used animal track which closely followed the dry stream bed. The deep shade, the craggy contours of the rocks and ledges, the large trees, the abundance of birdlife and the feeling of natural richness were a tremendous relief after the burning plateau. It was difficult to conceive how two such different places could be found so close together.

We followed the animal track for about two hundred yards, where we discovered our gully led into a much wider valley, just as shady and green. When we stopped I could hear the musical sound of running water. A few yards farther on, dipping and splashing around fern-covered rocks, ran a crystal-clear stream. In the trees directly above us were five fresh chimpanzee nests. At that moment I couldn't imagine paradise being any more inviting than this miraculous valley on the edge of the dry plateau.

We spent the day exploring, and I saw my first wild West African chimp—or rather his bottom as he fled into the vegetation at the sight of us. Much farther down the valley, where it opened out and became wider and less forested, we surprised a group of three elephants close to the water. That evening I concentrated on the vegetation, and recognized many trees I knew to have edible leaves or fruits. Already I was imagining Pooh and William here, meeting other chimps, beginning a new life. There was more of the valley to investigate, not to mention the rest of Mount Asserik. Perhaps I would find many such valleys—some even better than this one, if that were possible. I tried not to be impulsive but I felt happy and exalted within myself. This valley might prove to be the promised land.

An exciting few weeks followed. As I suspected, there were many other forested valleys which held signs of wild chimpanzees, but I discovered only one that contained a running stream, and it was on the other side of the mountain. From my limited knowledge it seemed to me that plant food for the chimps, as well as innumerable baboons, vervets and patas, was plentiful. Several times I glimpsed wild chimpanzees, and each time I hoped that Tina, Cheetah or Albert might be among them. I was never able to get close enough to the chimps to study their faces, and if any of my three were with them they gave no indication.

By the end of February I had explored most of Mount Asserick on foot and had discovered no place more suitable as a camp, or rehabilitation site, than the valley we had come upon on the first day. Though I had still found no sign of Tina, Cheetah and Albert, I felt I could not afford to spend any more time searching for them. As March approached, I began to get impatient to start the rehabilitation of William and Pooh. Having them with me might even facilitate getting to know the wild chimps and finding the others. At the end of the first week in March I packed and set off on the long journey to The Gambia.

While I had been away, William had become more and more difficult to handle during the morning walks, and Nigel had had many problems trying to retrieve all that he stole from the visitors. Finally it became necessary to keep William and Pooh in the enclosure whenever a large party of people was expected, and Nigel would take them for a walk later in the day. William was also difficult to get back into the enclosure after the walks, and would either keep dodging round the rest house or go into the hyena pen and play with the two hyenas.

It was worrying when he decided to tease the antelope orphans, for he caused them to panic and run into the wire, damaging their faces. Also, on several occasions he entered a hutch of guinea fowl which my father was proudly rearing. They were about half-grown and almost ready for release. When William became bored with chasing them featherless, he would catch, kill and attempt to eat them. All in all, William had become a tyrant and was exasperating everyone who had anything to do with him.

The evening before Nigel and I were due to take the chimps to Niokolo, William escaped from the enclosure. Having played with the hyenas, he entered the serval cat enclosure and killed a young kitten before I could stop him. Daddy was too disappointed and sorry for the kitten to be angry with William, but I think the episode helped a great deal to make the parting easier the following morning.

William and Pooh were used to riding in the Land Rover and thus didn't need to travel in a cage. We piled the back of the vehicle with cushions and foam mattresses and the chimps traveled comfortably.

All went well till we were almost back at the valley and decided to camp for the night. Near the car we met a troop of baboons. William bristled till every hair on his body was erect, and gave chase. However, when he realized that there were many more baboons present than we'd been able to see at first, he sat on a rock and observed them. Pooh, having watched his big brother acting so brave, and being blissfully ignorant that baboons could be dangerous, dashed forward, slapping trees and stamping his feet, his coat also bristling to its maximum. A large male baboon made a quick, aggressive charge in his direction which sent him running for all he was worth to the safety of my arms. Pooh had begun to learn that his new home was not as safe and secure as Abuko. That evening Pooh and William slept on their cushions close to our beds.

Next morning William had recovered his courage. It was the sound of his screaming that jolted me out of bed. Nigel and I raced over and found William about to take on an entire troop of baboons alone, confronting several adults which kept making short charges at him. As I ran forward the baboons scattered. William took heart from my proximity and began to chase them once more. I was sure that any moment he was going to be the victim of a mass attack, but I eventually caught up with him and carried him back to the car.

Living out of a Land Rover with two chimps constantly present can have its problems. William was forever stealing from the kitchen, and Pooh had a mania for rolling in blankets or taking them into inaccessible trees and leaving them there, folded over the branches. That morning William emptied our last jerry can of water, and Nigel and I decided that our first priority at Mount Asserik must be to organize a simple camp.

With mounting excitement I drove toward the mountain and the place which I already thought of as my valley. Under the huge fig tree to the left of my earlier campsite, I found a few fallen fruits and some very fresh chimp dung. William stooped and sniffed it, then took a step backward, sat down and looked intently down through the trees to the bottom of the valley. I followed his gaze but could see nothing.

We quickly hurried out to explore the valley. We saw plenty of fresh buffalo and elephant dung but no more signs of chimpanzees. On the way back Pooh was in the lead. As we climbed across a fallen tree that barred the path, he suddenly jumped and gave the low alarm hoot. I picked him up and cautiously looked over the tree trunk. I saw about ten feet of very bloated python. The snake had obviously just fed and was feeling sluggish and lazy. I moved away fast. We hoped William wouldn't see the python, but as he crossed the fallen tree he gave a loud whaa bark, picked up a heavy piece of wood and threw it at the snake. The log landed a few inches from the snake's head and caused it to move off along the bank of the stream. William, hair bristling, approached to within about two feet of the snake, picked up a dead branch and hit it hard on the back.

The snake drew back instantly into a strike position and its head followed William's every move. I feigned dreadful fear, but William ignored all my warnings. Nigel edged forward, seized William's arm and dragged him away.

That evening it was difficult getting the chimps to settle down for the night—both refused to have their cushions in trees and insisted on bringing them to the ground to sleep. Lying on my camp bed, I looked up into the branches of a huge netto tree. Before I slept, I chose two parallel branches on which to build a sleeping platform for William and Pooh. I was sure that once they got into the habit of sleeping high up in a tree, they would no longer feel secure on the ground.

16. REUNION

William and Pooh woke us at dawn: Pooh by trying to get into bed with me, William by hammering on the Land Rover to break into the food supplies.

We discussed the platform I'd thought about, and decided to build it immediately. As we were talking we suddenly saw, on the other side of the gully, a group of wild chimpanzees filing down to the stream. I could make out four adults with infants clinging to their bellies, two juveniles and another two adult chimps bringing up the rear. As quietly as I could I picked up Pooh and pointed toward the wild chimps, but by the time he understood that I was trying to show him something, the strange chimps were already in the trees and much less obvious.

Nigel and I hurriedly led William and Pooh to a spot farther upstream and waited, hoping to intercept the wild chimps. We had been seated there quietly about a quarter of an hour when a juvenile chimp appeared on the edge of the plateau and began to descend toward the stream. Nigel and I froze but Pooh and William continued to move and act normally, and the wild chimp spotted them almost immediately. It stopped, curious but seemingly unafraid, and looked at us all intently. Seconds later an adult female with an infant appeared behind the juvenile. As soon as she saw us she turned and disappeared without a sound onto the plateau. The juvenile followed.

Neither William nor Pooh had noticed the visitors. I was anxious not to alarm or frighten them in any way, so rather than continuing to follow the wild chimps, I led Pooh and William back to camp and hoped that it would not be long before the next opportunity arose to introduce them.

Before we had settled in, the news came that Tina had been observed several times in the area of Simenti, fifty miles away by road. She had been seen alone, frequently on the periphery of a troop of baboons. Sometimes she moved around, covering remarkable distances in a day. She was relatively unafraid of cars and used the park roads a great deal when traveling.

We repacked the Land Rover in record time and by evening were near Simenti, overlooking the River Gambia. William and Pooh, despite the long journey, were full of mischief. They were in and out of the car, stealing, eating and creating near-chaos, before I was finally able to persuade them to sleep.

Nigel and I were awakened by the chimps at dawn. Not long after we'd got up, a safari bus from Simenti Hotel arrived. I asked the driver when he had last seen the female chimpanzee. He told me his clients had watched her for several minutes only the previous evening. She had been on the road to Patte d'Oie. I relayed all this to Nigel. Patte d'Oie was just three or four miles from where we were. "Do you think we'll find her today?" I asked. "I hope so," Nigel said gently. "I can't think of a more appropriate birthday present for you." At that he handed me a rather grubby envelope and wished me a very happy birthday. I'd forgotten the date and stood speechless. With his usual foresight Nigel had brought the card all the way from Gambia, and despite all the excitement he had not forgotten to deliver it at the right time.

As we approached the place where Tina had last been seen, we found a set of well-defined chimp tracks leading down the middle of the sandy road. We followed these for three and a half miles before they turned off the road into a small forested depression. I stopped the Land Rover and we got out. There were no tracks now to follow, so we set off at random. Barely five minutes later we reached a thicket of bamboo. As we entered it Nigel froze. "Did you hear that?" he whispered. I shook my head. "A chimp, the soft 'oooh' sound they make when alarmed. It came from over there."

We came to a wide, dry waterway. As I was scrabbling up onto the other side, I saw her. Not twenty yards away among the dry bamboo stood a large chimp. Every hair on her body was erect and she was swaying slightly from side to side, a slim stick of bamboo in her hand. Unusually bright orange eyes glared past me at William.

She looked splendid. William stood frozen in his tracks, Pooh similarly immobile just behind him. William reached out, his coat erect and bristling, took a branch of a small shrub in his hand and shook it vigorously. For a few seconds they faced each other.

Still upright, William advanced toward Tina, slowly at first but with gathering speed. Tina stood her ground until William picked up a piece of dead wood and flung it at her. She dropped to all fours and dodged it, ran for a few yards, then stopped, screaming harshly. She was in estrus and sat down on her pink swelling as William approached her. The screaming changed to a continuous hoarse panting as he neared her, and William, walking now on all fours, his coat sleeker, replied with the same panted greetings.

Bobbing slightly, Tina extended the back of her wrist to William's face in a friendly, submissive gesture. William took her wrist in his mouth, at which Tina immediately swung round and presented her backside to be mated. William, still panting loudly, didn't hesitate a second but mated Tina normally. Pooh rushed over, hooting, and laid his hand on Tina's shoulder. She turned, a wide grin on her face, and making small squeaking sounds began to groom William frantically. Pooh, caught up in the excitement, began in his turn to groom Tina. They remained thus for a few minutes while I pinched myself to make sure I wasn't dreaming.

The three chimps groomed each other for about fifteen minutes, then William mated Tina again. She led the way to a large netto tree in fruit, and William and Pooh followed her into it. Since our arrival in the park the only wild food I had seen William and Pooh feed on had been kenno leaves. Now, having watched Tina eat for a few minutes, both of them began to chew netto pods with as much relish and food-grunting as they ate mangoes or bread and jam.

Tina had given no indication that she recognized me, so I remained at a distance and observed, anxious not to disturb the freshly reestablished comradeship. Apart from being unafraid of human beings, Tina was now a wild, independent creature. I wondered what had become of Albert, why he was no longer with her, whether I would ever see Cheetah again. Not for the last time I wished that Tina could speak and tell me her story.

I had found Tina, but what to do next? Except for a grizzled old male chimp that paid short, annual visits to Simenti, wild chimpanzees were rarely seen in the area. The sparse, scrublike vegetation and flat terrain were unsuitable for them. It provided ideal country for antelope, however, and consequently the density of large predators such as lion and leopard was higher here than in other parts of the park. For this reason, and because of the comfortable hotel in the nearby town of Simenti, it was the area most frequented by visitors to Niokolo Koba Park. Moving the rehabilitation camp to Simenti was unthinkable, so I was left with the alternative of taking Tina to Mount Asserik.

Tina had been free for two years and had coped admirably. I wondered whether it was unfair to interfere in her life again. However, when I considered that she was entirely alone, and how much she seemed to appreciate William and Pooh's company, I felt justified interfering just once more to move her to the comparative lushness of Mount Asserik, with its abundance of food and company of her own kind.

The problem of how to transport her was quickly solved. Soon after my arrival in Senegal I had met Claude Lucazan, an enormously strong and stocky man with bright blue eyes in a sunburned face. It was clear at our first meeting that he considered me mentally deficient; it quickly became equally clear that he would spare no effort to save me from the consequences of my insanity. He had already been of immeasurable help to me and now volunteered to lend his trailer as a traveling cage. It would, however, be a couple of weeks before the trailer would be available, so we settled down to make the best of things in the meantime. One problem proved to be our water supply. In the roots of a nearby tree was a small and murky water hole, about the size of an average water basin. Even if we had enjoyed its exclusive use we would soon have run short; as it was, the water hole was crowded with thirsty bees for most of the day, and we were forced to rely on the Simenti Hotel, some fifteen miles away.

Tina was the only one among us who was prepared to brave the bees, so the drinking hole became her private preserve. I tried to persuade William and Pooh to drink there during the evening, when the bees were less numerous, but they both refused. Perhaps the location of the water and its color were too much for their still refined tastes.

The next few days were spent following Tina, William and Pooh around, and taking notes. Each day the relationship between them seemed more firmly cemented. Tina was exceptionally possessive of them both. While her swelling lasted, William tried to make up for his bachelor existence during the past two years and mated Tina at every possible opportunity. I did not see Pooh try. He shared more of a platonic friendship with Tina and spent hours playing around or with her. Tina occasionally tickled and mothered him, but spent a great deal of her time grooming them both. She was so anxious to keep them with her at first that she resented humans approaching too closely. Several times she charged at me aggressively. Fortunately I always managed to avoid a confrontation, but once she caught Nigel's ankle, neatly flipped him over and bit his foot.

Tina also considered it her role to protect the younger chimps from any attack or criticism. The moment I raised my voice or threatened either of them, she would begin to display. Hair erect, she would swing branches back and forth, then leap to the ground and charge off through the dry, crackling leaves. William and Pooh soon learned that if Tina was close by, they could far more easily steal food and misbehave in other ways. I hesitated to admonish them in front of her in case I should upset or frighten her. Tina's ties to the camp were still frail, and I was anxious to keep her with us until we could move her. During the first few days, every time Tina was missing I would worry that perhaps she had continued on and left us. During her short absences I found it difficult to concentrate on anything else. It was always with intense relief that I'd listen to Pooh or William's greeting pants as they spotted her again. To cut down her wandering, we fed her as much bread and as many mangoes and biscuits as she wanted.

While I spent most of my time with the chimps, Nigel constructed a bamboo feeding table. Each day a few more upright bamboos were added, to gradually enclose the table on three sides. We hoped that if Tina learned to eat in this semi-enclosed space, she would enter the trailer cage more easily. She was noticeably nervous about it at first, which made me pessimistic about her acceptance of the trailer, but seeing that William and Pooh were completely uninhibited, she began to relax and even climbed onto it to feed with the others.

Tina had filled out a lot in the past two years. She was now a large eleven-year-old female: her head had broadened and she had become balder. Her eyes seemed smaller, and she had reverted to the original mistrust of people she had shown on arrival at Abuko. But there were certain characteristics I recognized. She still had unusually orange-brown eyes and a few of the old mannerisms, the most distinctive of which was an affected-looking twitch of the lip. I felt sure she remembered William and Pooh and that it was not just a desire for company of her own kind that kept her so close to them.

We were anxious that Pooh and William should spend as much time as possible with Tina. Often during the day she would leave the immediate vicinity of the camp to search for food. Pooh and William would follow her only about a hundred yards out of camp unless we were present; if they were by themselves they would turn back and wait in camp till she returned. Whenever we could, we would follow the three chimps, keeping about thirty yards behind them so as not to upset Tina, but close enough to reassure Pooh and William. Pooh would quite often prefer to walk with us; William usually remained near Tina.

One afternoon, on one of these walks, we lost sight of William and Tina for a few minutes as the vegetation ahead became denser. I had been aware that a troop of baboons was somewhere ahead of us, for one or two of them had heard us approaching and had begun barking. Suddenly there was a terrific commotion. Through all the baboon sounds we could hear William screaming and Tina whaaing and screaming alternately. We ran forward and met William coming toward us in full flight. By this time the baboons were in sight, but they fled as we approached. Tina, every hair on her body erect and a large chunk of wood in her hand, pursued them for a short distance. It was difficult to keep Pooh from following her example. William, we hoped, had learned his lesson, for he had superficial bites on his hands and feet, several long, deep scratches on his left arm and two deep canine punctures on his right arm. While we examined William, Tina continued to make charge after charge in the direction of the baboons, throwing branches and leaves at them. The baboons kept on barking but made no attempt to attack Tina, possibly because we were close by.

Almost every evening Tina made a fresh nest. Only occasionally did she use the same one for two nights running. If she chose a site close to camp, I'd try to make sure that William and Pooh watched her at work. The evening the baboons attacked William, she made her nest in a large tree above the Land Rover. William and Pooh watched, and then, instead of going to bed on top of the Land Rover as usual, William climbed up and sat on a branch near where Tina lay comfortably stretched out in her nest. He went to inspect a nest she had made a few nights previously, and after some hesitation climbed into it and lay down. I was thrilled. This was yet another example of how Tina's presence was helping William and Pooh.

WHEN WE HAD BEEN at the Simenti camp for 2 weeks, Claude returned, towing a small zebra-striped trailer with iron mesh sides and roof behind his car. It was an ideal cage for transporting Tina. We wired the door of the trailer open and began to feed the chimps from it. Tina was highly suspicious at first, and for an entire day would go nowhere near the trailer. Gradually, through watching Pooh and William feed and play in it safely, she began to lose her fear, and at the end of the second day she was entering the trailer briefly to snatch a few mangoes.

On the morning of the third day we rigged up the door of the trailer with a long cord, enabling us to slam it shut from a distance. I placed a pile of mangoes and biscuits at the far end of the trailer and waited. Pooh and William were in the trailer the moment they woke up, and Tina, still timid, followed quickly and grabbed a few. Unfortunately, William was sitting in the doorway with his legs dangling over the edge, so we couldn't close the door. I had to keep replenishing the food stocks in the trailer till finally the opportunity arose to trap Tina.

Tina went wild when she found she was locked in, and I was afraid that she would hurt herself. We piled Pooh and William into the car and Nigel and I set off. Once we were moving, Tina became quiet, sitting in the trailer and looking around at the passing landscape. We were almost halfway to Niokolo when one of the tires blew out. William and Pooh sat on the trailer while Nigel and I changed the wheel. To my relief Tina did not start throwing herself around again but sat quietly watching us.

We reached Mount Asserik just before midday. I was worried that after suffering such indignities Tina would run away, never to be seen again; but though she ran out of the trailer, she soon slowed to a walk, climbed a laden fig tree in the corner of camp and fed. William and Pooh joined her. That afternoon Tina was missing for several hours, and my fears returned, but she was back that evening to nest in one of the trees on the edge of the valley. She has remained with us ever since.

PART THREE

LEARNING TO BE FREE

17. SETTING UP CAMP

Upon returning to Mount Asserik with William, Pooh and Tina, I employed Charra, a tall young man of the Besseri tribe, to help me in camp. With Nigel about to return to Abuko, some assistance was clearly going to be essential. The night before he left, he and Charra helped me organize a camp. We cleaned out Tina's trailer and stacked it with the food supplies. A small orange tent we'd brought served as a store for jerry cans and basins and the Land Rover made a passable wardrobe. Charra chose a site for the kitchen and constructed a small fireplace with a crude rock shelf next to it. The big tin trunk was wired to a small tree nearby, and Charra filled it with pots, pans and the cooking ingredients he regularly used. That evening we slept on camp beds grouped around the netto tree. William and Pooh were on the Land Rover's roof rack, while Tina made a nest in a tree at the bottom of the gully. I slept soundly that night, content that Tina was at Mount Asserik and the project had really begun.

The following morning Nigel left. I had come to rely heavily on his help and companionship during the past weeks, and I felt dreadfully alone. Charra spent the day finishing the platform Nigel and I had started before leaving for Simenti, and I helped him where I could. William and Pooh spent many hours on the edge of the gully, playing, feeding on figs or resting with Tina.

By that evening the platform was finished, the crosspieces of wood were securely lashed down with rope, bark and wire and the rungs to the ladder had been similarly lashed on to the uprights. The platform was about thirty-five feet high and looked like the beginning of the Swiss Family Robinson's tree house. William and Pooh had visited it at intervals during construction, but Charra had

The nesting platform

impatiently shooed them away as they tried to steal the rope and fling down pieces of wood, or carry them to different branches inaccessible to humans, where they would try to make their own platforms.

William had spent a good half hour thoroughly examining the trailer to see if there was any means of entering it to get at the food. I had packed the trailer so that all bottles and supplies in packets or plastic bags were in the center, well out of reach of prying fingers that could be pushed through the mesh. William spotted this at once. Having tried every possible combination to tug, push or lever the door open, he found a long, slim piece of dry wood with which he was able to prod the flour sack and drag soup packets within reach of his fingers. By the time I realized what he was up to, he already had a packet of soup in his hands. As I ran toward the trailer, he scurried off into the gully, peeking at me insolently from under his arm. It was the beginning of a battle of wits over the protection of my possessions, and it was to be a long time before I won any of the bouts.

That evening, when Tina went off and set to work building a nest, I began to praise her like mad. William and Pooh sat beside me and kept looking from me to Tina. "What a clever girl, Tina! Oh, Tina, what a clever girl!" I kept repeating. William and Pooh loved being praised, especially by people they were fond of, and I hoped that if I praised Tina every time I saw her make a nest, it would encourage William and Pooh to build one too. At least it focused their attention on what she was doing.

I placed Pooh and William's cushions up on the platform with a bunch of leaves and sat with them till it was almost dark. They both stretched out, and I was hopeful that they would stay. However, five minutes after I had descended I found Pooh sitting on the camp bed beside me. I tried to put him up on the platform again, but each time I left, William pushed Pooh down and stole his cushion. Finally I was obliged to make another bed for Pooh on top of the Land Rover.

In the next few days the chimps and I began to go for walks in the valley near the camp. On these first walks I could never really relax, but I had to pretend to, for if William and Pooh sensed I was afraid, they would become uneasy too and creep around quietly rather than enjoy themselves and get to know the new environment.

There was an abundance of elephant and buffalo dung along the stream, and my ears were continually pricked for the sound of animals approaching. During one of our first walks I sat beneath a tree near the stream. The vegetation is thick in the valley and it is always shady and dank there, even at the end of the dry season. I was watching Pooh and William feed when there was a loud call directly above me. I almost jumped out of my skin, but as the sound was repeated I realized it was a large brown bird with a long neck—the hadada, as I later learned—winging its way heavily above the stream. The sudden cries of a fish eagle or other predatory birds were continually startling me at first, but I gradually became accustomed to the different sounds till they became as much a part of my day as the fire crackling or Pooh and William panting.

Each evening William and Pooh would watch Tina nesting, but neither of them showed any signs of wanting to do the same. William generally slept on the platform and Pooh on top of the Land Rover. After supper I would light an oil lamp, write up some notes, read a little, then try to sleep. One night, however, soon after our arrival, was less peaceful. Charra had discovered a group of poachers in the vicinity, hunting for buffalo or, worse still, elephant. He had gone off to Niokolo to warn the park warden but had not reappeared by nightfall, nor was there any sign of him when I woke just before dawn. I was quite worried by this time. Perhaps Charra had fallen, sprained an ankle, been attacked. Perhaps even now he was lying injured somewhere.

I decided that I would drive as fast as I dared to Niokolo to look for Charra and, if necessary, get help. I already had the car keys in my hand when I noticed that one of the back tires was completely flat. I changed the wheel, but there was no way I could get into the car without Pooh leaping in with me. I wondered whether William and Tina would stay alone. I drove off; Tina and William began to walk quickly and steadily in the tracks of the car, William stopping only to scream his distress at my deserting him. I drove back into camp and sat down to think. Finally I put Pooh in the stationary zebra-striped trailer, with Charra's wife, Nierri, who was staying with us at that time, acting as Pooh's somewhat reluctant companion. Then I scattered lots of biscuits and fruit around camp, and drove away quickly while William and Tina's attention was distracted.

Exploring near camp

I was worried sick about leaving the chimps alone in camp and cursed the roads for being so rocky and potholed. An hour later, and about two miles from the guard camp I met Charra, walking back to Mount Asserik. He said there was no gasoline at Niokolo, so the guards could not use their Land Rovers. We drove to the guard camp, and I offered to take the guards with me if they would walk back. The offer was gratefully accepted, and the Land Rover staggered back to Mount Asserik carrying ten people—Charra, myself and eight armed guards.

I had been away exactly two and a half hours. William greeted the Land Rover noisily; Tina was on the edge of the gully. Nierri and Pooh were both still in the trailer, Nierri looking extremely uncomfortable. In our absence William had unscrewed the trailer legs and let them down unevenly. He had then rocked the trailer hard several times, scattering tins and supplies all over and covering Nierri and Pooh with flour. It was only by a miracle that William had not managed to turn the trailer over completely.

I kept the chimps occupied while Charra led the guards off to find the poachers' camp. Suddenly it sounded as if a war had broken out. A volley of gunshots rent the air. Pooh leaped into my arms, William grinned his fear and clambered into a tree and Tina vanished at high speed into the gully. I hoped that she would not desert us completely as a result of her fright.

It was a good hour later that I saw William freeze and then stand to peer through the scanty tree line onto the plateau beyond. I straightened from my squatting position and rearranged Pooh on my hip. He seemed to sense the tension and insisted on staying securely clutched to me. I walked over to William. Without shifting his concentrated stare from the plateau, he wrapped an arm round my thigh for comfort. I peered in the same direction, and after a few seconds was able to detect a line of uniformed men pushing a figure ahead of them. When the guards reached camp I discovered that, although the guards had moved quickly, all except one of the poachers had managed to escape. The one they had captured had been made to carry the putrefying head of a buffalo, which oozed maggots and stank so strongly that if I got close to it I found myself retching. The buffalo had been killed by the poachers. It had been a female, and inside her there had been a perfectly formed little calf.

The poacher showed shocked incredulity when he was pushed into the camp and found William, Pooh and myself there. He was forced to sit under a tree at a safe distance from camp, but never, it seemed, did he take his eyes off the chimps. For the first few moments William and Pooh showed distinct anxiety at the sight and smell of the fly-filled buffalo head and made several long high-pitched whoops at it. I had heard the sound only once before: when Tina had come across a dead and stinking crocodile near the lake at Abuko.

After breakfast next day I was intensely relieved to see Tina appear in the trees on the edge of the gully and sway at us all. William and Pooh raced over to her. Tina panted loudly as they approached, and the three of them began a frenzied grooming bout before climbing the fig tree to feed on the fruit that remained.

A FEW HOURS AFTER TINA'S RETURN, I heard Charra calling me in a stage whisper. I looked up into the netto tree to see him pointing frenziedly at the small plateau on the west side of camp. "Wild chimps," he said urgently, "wild chimps!" Through the tree line I could see a string of chimpanzees, following the same path as before. A female was leading, with a young infant clinging to her stomach and a juvenile walking beside her. A second female followed with an infant perched jockey-style on her back, and behind her was another adult chimpanzee with no infant.

It was obvious that my chimps hadn't seen them. As quietly as possible I led Pooh, William and Tina down through the gully to the stream. I was hoping that the wild chimps were going there to drink, as they had done on the previous occasion.

We waited a quarter of an hour and nothing happened. I tried to remain concealed, but Pooh was playing by himself, energetically swinging about in the low branches and generally making a lot of noise. When I concluded the wild chimps were not coming to the pool, I tried to lead the chimps upstream. About a hundred yards farther on I caught a glimpse of an adult chimp near the top of the slope. It had seen us and was hurrying on to the plateau. Again, none of my chimps noticed it. I climbed up to where I'd watched the wild chimp, but there was nothing to be seen. We searched

about for another thirty minutes, then gave up and returned to the stream bed.

I was so preoccupied with the peacefulness of my surroundings that I did not see Pooh steal the soap I'd brought. By the time I noticed, he was up to his ankles in the shallows, lathering his head and face, licking the abundance of foam off his fingers. I spoke to him crossly and told him to drop the soap. He slipped the bar into the water, but not before he had taken a substantial bite out of it, and ran off to a calmer part of the stream. William chased him. Pooh screamed and dropped his remaining morsel of soap, which William retrieved. William then entered the water next to Pooh and began studiously to lather his leg. Tina approached and watched closely. Gingerly she reached out and took a blob of foam from Pooh's head. She sniffed it several times, looked, sniffed again, then tasted it, only to spit it out immediately and shake the remains of the fast-vanishing froth from her finger. She wiped her finger carefully on a dead branch and some leaves. Tina, it seemed, did not share William and Pooh's taste for soap.

I FOUND I WAS BEGINNING to relax much more. I was not so constantly on the alert for sounds and movements. I was still aware, but aware of everything, not just the dangers. Aware of the peace, the strange beauty of the place, the dappled shapes made by the sun through the leaves; aware of the colors and distortions of the heat haze hovering over the bare rocky plateaus, of my own small place in this kingdom ruled so differently from the one I'd come from. Here, everything seemed organized and fitting. My peace of mind wasn't something I felt I'd achieved on my own but was simply absorbed from my surroundings.

We finished lunch about four thirty in the afternoon of the day we sighted the wild chimps. I set out to walk to the Land Rover. Pooh whimpered loudly and ran toward me, quickly climbing onto my back. William and Tina were also beginning to follow. I did not want the chimps to see me walking in the middle of the large open

Pooh (top) and William (bottom) loved to play with soap.

plateau which was my shortest path. I wanted them to be wary of open spaces, to walk cautiously if they ever had to cross one and to keep to the tree line whenever they could. The chimps would be extremely vulnerable to predators in such barren places. Instead of cutting across the plain, I followed the edge of the tree line, and when I finally had to cross open ground to get to the car, I exaggerated the nervousness and caution of every movement. I was gratified to see that Pooh and William understood and also looked around and moved cautiously. Tina refused to follow us at all. She had not felt safe in the open, and not having the blind faith in me that Pooh and William had, sat where she was and waited.

Back in camp that evening, I heard Tina begin to build a nest and dutifully led the chimps over. Pooh followed and sat on my lap, listening to my praise of Tina and watching her build her nest. I looked round for William, but he was not in sight. When Tina had finished her nest and settled down, grunting good nights to herself and the world, I walked back to camp. William was arranging a length of nylon rope around himself with one hand while he clung to a spade with the other.

Both articles had come from the Land Rover. Knowing that all the doors were securely locked, I walked to the car in some perplexity. William coughed at me several times as I passed him, but since I did not respond he continued his activities with the rope and spade. All the doors were still locked, but one of the small oval windows on the side of the back door had been pushed in. It wasn't broken and the thick rubber trim was still intact. William had taken what he could reach through the opening. I patched up the hole with a crisscross of thick plastic-coated wire.

I fed William and Pooh with a mug of warm tea and a small plateful of rice and sauce, which were the remains of lunch. During the day Charra had constructed a second sleeping platform a few feet away from the first. Almost without having to be told, William went straight to this platform and lay down on his ready-made "nest." Charra then brought up Pooh's cushion and blanket and a bunch of leaves and put them on the first platform. William watched idly as we prepared Pooh's bed but then rolled over and appeared no longer interested. Pooh seemed to like his bed, but I remained with him till he lay down and began to doze, then quietly descended the ladder. To my great relief he didn't follow me but stayed on the platform.

I lit two lamps, wrote a little more in my notebook and then took out my bed and began to make it. Charra had just announced that supper was ready when a sleepy Pooh pushed his way onto my lap and snuggled into my shoulder. I spoke to him for a few minutes, then carried him back up the ladder to the first platform. William coughed and grunted at me as my face came level with the platform —a sign that he knew I would not be pleased.

"Go on, you big bully!" I said. "Back to your own bed and let Pooh sleep!"

I spoke jauntily but with an underlying firmness which indicated that I expected him to do as I asked. William sat up; he too looked sleepy. I gave him a quick hug, then guided him off the platform onto the thick branch that supported it. William moved a few yards, then sat down. "Go on, William," I urged. This time there was only a trace of playfulness in my voice. William looked at me and began to climb toward his own platform. "Well done, Willie boy," I said gently. "Well done, thank you!" He grunted halfheartedly and rolled onto his cushion. Pooh settled fairly easily, but again I had to wait till drowsiness numbed his senses before I could move away.

NEXT DAY CAME A STILL FIERCER confrontation with the baboons. Breakfast was over and I was preparing for another day in the bush when suddenly there was a tremendous bout of noise—baboons crying and barking and a chimp screaming wildly. I imagined Pooh or William being torn limb from limb by a group of irate baboons.

Charra got to the scene before me. As I arrived behind him, the baboons were retreating, with William in pursuit hurling large rocks and branches. Pooh was fighting for the courage to back up William. Whaaing hysterically, he followed his big brother bravely—as long as it was just baboon backsides that he saw. The moment one turned to face him, he stopped, retreated and whaaed even more loudly than before. Tina added an occasional burst of encouragement but had climbed onto a low branch.

Charra explained that he had seen Tina catch a young baboon. Several other baboons had mobbed her, and in the end she had let go of her prey to defend herself. As soon as the baboons had perceived Charra and me, they had begun to withdraw, but before our arrival they had been returning the attack of the chimps. Tina did not seem to have been bitten, and when I finally managed to retrieve Pooh and William, they were unharmed too. I felt out of breath and a little shaken by the incident, but Pooh and William chased and tumbled about with each other in extra-high spirits. I hoped that the situation would not be repeated each time we encountered baboons.

That night was the start of the rainy season. Claude Lucazan had arrived providentially with two Senegalese assistants as the thunder began to roll, and had announced that he was going to build me a shelter. I was outraged:

"Claude, wait, I don't want to cut a single tree. I don't want to spoil this beautiful wilderness with permanent huts and houses. I don't want to destroy anything. You don't understand, I feel an alien here—small and insignificant. I'm overawed and humbled. I have no right to do it. Please, Claude, I don't want any buildings here. I can manage. I can manage, I assure you!"

Claude ignored me. Considering the trouble and probably expense he was going to in order to see me reasonably comfortable, I found it difficult to protest strongly, but even if I'd screamed it would have made no difference. Claude knew that some sort of shelter was necessary whether I liked it or not, and that once the rains really began, I would understand why.

It was certain to rain during the night, and finding shelter for everyone was a problem. The orange tent was only big enough for Charra. Claude's two assistants slept in my Land Rover. Claude placed two poles against his Land Rover and set a tarpaulin over them to form a ramshackle lean-to, beneath which we put our camp beds.

I was awakened at about 2 a.m. by the force of a tremendous wind. The little orange tent was flapping wildly and the trees all around camp were bending and swirling in the gale. I went outside the lean-to to see how William and Pooh were reacting. I was soaked within seconds. Poor little mites, I thought, they'll be dripping, but they must get used to the rain. Thunder and lightning they knew, but at Abuko there had always been shelter from the rain.

The rain was so heavy that even when lightning illuminated the camp I could not see the platforms. Suddenly through the sounds of the storm I heard Pooh begin to scream. I ran to the ladder and met him at the bottom. As soon as he was safely in my arms he stopped crying and buried his face in my neck. William was also coming down the ladder. His growing independence forgotten for a moment, he clung to me too. The fact that I was not afraid calmed them both. We sat embracing each other at the base of the netto tree and witnessed the storm together. I talked to them cheerfully through chattering teeth, trying to convince them that there was nothing to fear except the discomfort of being wet and cold.

After a quarter of an hour I was shivering uncontrollably and felt very cold. William and Pooh kept shuddering spasmodically too. When Claude called me, shouting that I should get out of the wet, the temptation was too great. Chatting as brightly as I could through locked jaws, I carried William and Pooh to the lean-to. Claude gave me a towel and I rubbed them both as dry as I could, then changed out of my wet clothes. Pooh was already under the covers and William was lying at the bottom of the camp bed. There was not really enough room for me, but we managed. All three of us slept comfortably till the morning.

It was gray and still drizzling when I awoke, and my bed was plastered with mud from the midnight adventure. Brown-smeared chimp prints seemed to be everywhere. After the dust and drought of the dry season, it was almost possible to watch things revive and grow before our eyes, and there was a beautiful smell of damp earth and cleanness in the air. Pooh and William's coats looked fluffy but shiny and healthy after their involuntary bath.

As the sky cleared, William climbed into a small tree and made a crude nest. I was delighted and wanted to tell him so, but each time I took too much notice of his efforts, he stopped and almost descended the tree. I got the distinct impression that he wanted me not to look at him, so I pretended to do other things. Finally he lay down to sleep and dry off in the sun in the same way that Tina, Cheetah and Albert had done.

I was certain that William understood the rudiments of nest-building. He may not have been good at it, but he knew what to do. My problem was persuading him to do it. William was a very contrary creature. He much preferred to accomplish what he knew to be forbidden than learn what was considered praiseworthy. Pooh would do things just to please me; William was reluctant to "work" unless there was a tempting reward for his labor. The older and more independent he got, the less he was inclined to do me a favor. Sometimes I felt that if I tried to stop him building nests, he would learn to be proficient at it twice as quickly.

While carefully paying no attention to his efforts, I gave him some surreptitious help by piling leaves onto his platform and then putting a branch in such a position that he had to arrange some sort of nest before he could settle comfortably. The use of the platform was, of course, a temporary measure; when I thought he was experienced enough at making nests and had acquired a strong preference for sleeping at a height, I planned to take it away. I hoped that then he would naturally go to a tree, climb it and make a nest.

Claude and the men began work early and by the end of the day a skeleton of the shack had been erected. Pooh and William had been picking up and attempting to use the tools, climbing on and off the structure, testing its strength as we went along. Though they were reasonably good at handing things back when they were needed, they slowed down the work so I took them for a walk. By the end of the weekend my wooden shack was built. It wasn't as large as Claude had wanted it to be—I had made him halve the original size he had mapped out—but it was quite large enough for my purposes. A big blue tarpaulin was stretched over the roof and covered with woven mats for camouflage. Claude took the measurements for a door and said he would bring one with him as soon as possible, but at least I could get out of the rain. He also planned to bring wire mesh to cover the windows so that the light could enter but the chimps could not.

The following week it rained several times during the day. Tina's calm attitude to the storms helped William and Pooh accept them as periods of discomfort rather than danger. Pooh frequently made ground nests during the day, using leaves of shrubs or dry grass and arranging everything in a circle around him. After one of the storms he climbed into the trees above the shack and actually began to pull branches toward him. He worked for about five minutes, and by the time he had finished had constructed a reasonable nest. At least I was able to praise Pooh for his efforts without risking any ill effect. It was a great moment. I knew now that both of them were able, if they wished, to build a nest.

Perhaps because of my apparent approval, perhaps through the excitement of achievement, a few hours later that morning Pooh climbed high into the netto and made another, still more satisfactory nest. I hoped that I might be able to persuade him to sleep in one of them at night, but almost as soon as he had completed the nests, Pooh began to play in them and more or less destroyed the evidence of his accomplishment.

18. WILLIAM GROWS UP

During the days we spent exploring the area, Tina helped to point out some of the other animals with whom we shared Mount Asserik. She also showed William and Pooh which ones were harmless and which we had to avoid.

A snake caused her to stop suddenly, raise her hair slightly and make a detour. A large herd of buffalo was merely watched with interest, but when they caught my scent and began to mill around, she gave the low alarm hoot, which caused William to stop his brave advance and scrabble up the tree beside her. She was almost oblivious to warthogs, but was quite attentive to a herd of the larger antelopes, such as bubal or koba. Oribis and jackals, which we frequently saw on the plateau, were allowed to pass unheeded; also the smaller duiker antelopes, unless they happened to break cover close to the chimps, in which case they were chased with intent to capture. Red patas monkeys, usually seen on the plateau, were ignored, but the vervets, which preferred the forest, were stalked and hunted.

Several times as we walked back to the camp in the evenings we would hear elephants farther up the valley. William and Pooh showed all the signs of being afraid of the loud calls. Tina would look in the direction of the sounds but did not seem disturbed by them.

When I went out alone with the chimps, I tended to keep to the valley or to areas within a mile or two of camp. I felt uncomfortable and vulnerable outside this self-defined territory though I could just as easily have met with an accident in the valley as anywhere else. I was anxious though to know every part of Mount Asserik and to acquaint the chimps with as much of it as possible. It was not so much the fear of being attacked by animals that worried me as the fact that if I slipped and fell, sprained an ankle or broke a leg, and I was miles from camp with only Charra to look for me, I would be putting myself and the chimps in an unnecessarily dangerous position.

Charra came with us whenever possible. We would set off shortly after dawn and come back to camp early in the afternoon. In this way we began slowly to explore the areas farther away from camp. Tina would sometimes remain with us all day, sometimes suddenly disappear and rejoin us in camp the same evening or in the valley the following day. I preferred it when Tina was with us—she taught us all so much—and whenever possible we tried to keep to the terrain and vegetation she preferred.

AFTER ABOUT A MONTH of this existence, my father and Hugo van Lawick drove up to visit us. I think William and I recognized my father's Land Rover at the same instant. We flung ourselves into each other's arms and jumped around in uncontrollable delight. William was the first to leap onto the car, push through the window and embrace my father tightly—squeaking and grinning his intense pleasure and excitement. Daddy held him tightly, speaking and laughing to him and rubbing his back. Pooh's shy hug made pale greeting in comparison.

Hugo had come for the first of the few visits he was to pay in order to get the material he needed for the film. He was exceptionally patient and considerate, and spent hours following the chimps about with his heavy camera, until they did something interesting. It was extraordinarily bad luck that the most exciting event of the week took place on the only afternoon we left camp without the camera.

We had followed the stream down to an open patch of valley. Tina had chosen not to accompany us, so when we reached a fruiting tree, I stopped and pointed it out to Pooh and William, food-grunting to encourage them to feed. Both the chimps climbed. Hugo and I sat on a fallen tree. Small branches had sprouted vertically from the trunk and these provided shade. We realized that we must have been well camouflaged when a bush buck came to a pool literally yards from us and drank. Hugo and I froze, and it was not until William shook a branch at the bush buck that it took fright and fled. We sat quietly, wondering whether any other animals would come to drink. Pooh and William continued feeding.

Suddenly Hugo nudged my arm. I craned my neck to see what he was indicating. Something was moving, but I couldn't tell what. Suddenly the animals broke cover. I gripped Hugo's arm in excitement: wild chimps, three of them. At last, I thought, William and Pooh will see their first wild chimpanzees.

The chimps walked toward us, in clear view, along an open strip of ground that ran parallel to the stream. I sat with my stomach in a knot, waiting for the moment when Pooh and William would spot them. Suddenly I heard Pooh whimpering. Then he and William scrambled down their tree and came to Hugo and me. The wild chimps stopped on hearing the whimper. They saw Pooh and William and watched their progress down the tree till they reached Hugo. It seemed to take them an instant to believe their eyes—two young chimps running to two people! Then they turned and walked quickly up the slope. They were suspicious and wary, but not very afraid. The third, an adult male, kept stopping and peering at us beneath his protruding brows.

William and Pooh had their coats fluffed up with apprehension and excitement. "Go on, Willie," I urged. "Go on, go and meet them." As the chimps turned to walk up the slope, William left us and walked out into the open. The male stared at him. He also had his coat partially erect but made no sound or gesture. Pooh followed William cautiously. The two females had entered a clump of vegetation farther up the slope which was almost completely hidden from our view by a large boulder. The male followed them, and William and Pooh followed him. I did not dare move to get a better look in case I should disturb the meeting. I sat holding my breath and shaking slightly. I hoped the adult chimps would not hurt my younger ones. I very much wanted them to be friendly. I could neither see nor hear anything. All was calm.

Then a movement caught my eye. The three wild chimps were descending the slope again, with William behind them. Hugo and I were in plain sight. The chimps kept looking at us but walked calmly to the stream and disappeared among the trees. Pooh came straight back to us the moment the wild chimps began to move away, but William followed them to the stream, sat down for a moment or two watching them go, then returned to us. I was beside myself with excitement and relief. I looked at my watch. It was five past one and the chimps had first come into sight at three minutes past twelve, so William and Pooh had spent an hour in the presence of wild chimps and everything had been peaceful.

I WORRIED A GREAT DEAL about Tina. With all the activity about camp she had appeared less and less frequently, and had now been missing completely for several days. William missed her, I knew, for several times a day he would sit on the edge of the gully and stare down into the vegetation below, alert to every rustle.

Then Hugo left, my father having preceded him by several days. Julian—who had recently replaced Charra as my assistant—and I set out as usual with William and Pooh. We had gone only a few yards from camp when I saw William stand erect and, with arms held away from his body, begin to advance with short, mincing "fairy steps" toward the dry stream bed at the west side of camp. Expectantly I followed his gaze. Tina broke cover nearby, also standing upright with her coat fluffed out, and swaying slightly from side to side; then she approached William. She grinned submissively and squeaked her excitement as they met and embraced each other. Tina had a small pink swelling which she presented almost immediately to William. He responded by mating her, then both of them disappeared into the stream bed. Pooh rushed over and also disappeared.

By the time I found a spot where I could watch them without interfering too much with their activities, William and Tina were in a small fig tree grooming each other, and Pooh was feeding near them. When Tina saw me she panted and extended her hand to William's face, then presented to him again. William mated her once more. Pooh became excited and swung down closer to them. Tina promptly turned and presented to Pooh. Pooh hesitated, looking indecisive, but William, hair erect, advanced on Pooh and chased him away. Pooh fled, rather distressed, and having sat a few seconds rocking, flung himself into the neighboring tree with great bravado and began to play alone, swinging, free-falling and grabbing a lower branch.

Each time Pooh tried to entice Tina to play with him, William would accept the invitation for her, and play, rather boisterously, with Pooh. When he began to feed, if Pooh still insisted on trying to attract Tina's attention, William would shake branches angrily. Tina, grinning uneasily, almost always responded to William's branch-shaking by hurrying to him and presenting. Gradually, though, Tina became less responsive to William's gestures of annoyance, and he was obliged to begin a rough game with Pooh to distract him from getting close to Tina.

Later in the morning, when Tina began grooming Pooh, William ignored them for a few minutes and lay resting, looking out across the plateau. Then he suddenly sat up and began to sway the branch above them with increasingly powerful movements till I thought he was working himself up to attack Pooh. Tina glanced up at him quickly a few times but continued to groom Pooh, apparently undisturbed. William suddenly stopped swaying his branch and in desperation clapped his hands loudly. This time Tina jumped, then moved toward him. She didn't present but lay on a branch beside him. Pooh sensed that Tina was inhibiting William from an outright attack, and took advantage of the situation by continually approaching her with invitations to play. Tina responded by finger wrestling with Pooh, her mouth open in a chimpanzee play face.

William began to clap his hands and shake the branches, obviously distressed at the attention Pooh was receiving but unwilling to attack Pooh while Tina was more than likely to defend him. Finally Tina seemed to get tired of both her suitors, climbed out of the fig tree and disappeared. William followed her closely, and after only a few moments I lost sight of them in the vegetation.

William, now eight years old, was clearly adolescent. During the comparatively short time since we had left the reserve, he had steadily shown less need for affection. In Abuko he had looked for excuses to be picked up and cuddled. Now, not only did he refuse to allow anyone to carry him, but he shunned demonstrative affection, turning his face away if he had the slightest suspicion I might plant a kiss on it and firmly disentangling my arms if I gave him the usual good-night hug. I can only interpret his actions and expressions as embarrassment.

I knew that a wild chimp of his age would be starting to leave his mother for short periods to travel on the outskirts of all-male groups and gradually introduce himself into the hierarchy. That William was beginning to do the same pleased me, but at the same time I worried slightly. How capable was William of taking care of himself in this new home, which harbored so many more dangers than Abuko? I wasn't sure whether he was aware enough of those dangers to be safe without my protection. Then I shut off those thoughts. He was with a far more experienced lady than I, a far better teacher, and that was what we were here for after all.

Tina's swelling increased that night, and William's interest in her increased with it. The next morning they both disappeared again, but this time I continued the walk with Pooh, far more at ease in my mind about William's safety. Before midday I returned to camp. Pooh had fed and played most of the morning, and I had mail to answer. We found William sitting on the edge of camp with a small tin of milk in his hand and several packets of soup in each foot. He was either full or feeling decidedly guilty, or both, for he bobbed and coughed submissively as I approached him and handed me all but one packet of soup and the tin of milk. When I tried to take these he turned his shoulder, then got up and ran to Tina, who was sitting in the fig tree. She swayed menacingly at me as I approached them. It was obvious that should I attempt to force the issue, she would take William's part. As I was unwilling to take the risk of being attacked by an irate Tina, I had no alternative but to retire gracefully.

During the next few days I tried to keep up with the chimps, but it seemed that my presence disturbed them. Tina resented the attention Pooh gave me, and as soon as William realized there was no one in camp, he and Tina would sneak back to see what they could steal. Several times I only just got back to camp in time to stop William from completely destroying the tent. His strength seemed phenomenal for his size and, with his knowledge and experience of tool-using, he was an exceptionally successful thief. The door to the food trailer was secured at top and bottom with substantial padlocks.

One evening when I went to the trailer to take out supplies, I found it impossible to unlock. On investigation I found that William had been trying to pick the locks with a thin piece of wire, which had broken off in one of them. The other was full of shredded pieces of dry wood. William had obviously been trying to open it with twigs.

I was able to clean the padlock that had been jammed with wood, but the other proved such a problem that eventually we had to break it. The trailer door was made from half-inch iron and quarter-inch steel mesh. Apart from the two padlock attachments, it also had a stable-door latch at the center. The trailer had been

A big grin from William

minus one padlock for only three days when I arrived one evening to find that William had broken in. By swinging from the top of the trailer with one arm and repeatedly drumming on the door with the flat of his feet while his other hand manipulated the lever, he was able to unlatch it, despite the padlock below. Once unlatched, the door had a little play in it which he had exploited with his fingers till he had made a gap wide enough to insert a short but stout piece of bamboo. This he had expertly used as a lever and thus had bent out the top of the trailer door far enough to accommodate his long, muscular arms. His bamboo lever was still *in situ* when I discovered him, but it was only some time later, when the trailer was again safely padlocked at top and bottom, that I watched his methods of unlatching the door.

It had been my intention to discipline the chimps as little as possible once the project had begun, but I quickly found that if I was to survive in camp, William must be prevented from stealing all the food or running off with clothes and equipment.

In Abuko I had always been considered a supremely dominant being, but here I felt my control over William slowly slipping. I had to contend not only with his increasing age and size but also with the extra disadvantage of Tina's presence. I had to be extremely careful not to arouse her anger. Her two years of wild living had eroded much of the confidence toward humans that she had slowly acquired at Abuko, so she was far more easily intimidated than either William or Pooh. I felt that if she did attack me while defending William or Pooh, I could probably stop her by some sort of counterattack, but in doing so I might frighten her to a degree where she would lose the frail trust she had in me and run away. This was the last thing in the world I wanted. However, if I backed off or stood my ground passively, I was likely to be bowled over and bitten by ninety pounds of angry chimp. I went to extreme lengths to avoid having to choose between these two alternatives. William did little to help me in this respect. In a short time he realized that if Tina was present he could do as he pleased.

It was in one of these tense situations, with Tina sitting in a tree above the kitchen, that I watched William add another item to his growing list of surprises. I was sitting by the fire having a cup of coffee when the chimps came over from the tree where they had been feeding. The kettle was simmering on the fire and instant coffee, sugar and a cup with milk were sitting by the side of the canteen. William wanted my coffee. Looking at Tina, he placed his hand on my cup, slyly glancing back at me. He began to pull gently, continually staring at me. To have a chimp I had raised take advantage of me in this way made me feel really angry. I tightened my hold on my cup, opened my eyes wide, and in as deadly a voice as I could muster said: "Willie, don't you dare! Tina or no Tina, I promise you'll be sorry!" Almost immediately his hand and his eyes dropped. I felt he was annoyed, not so much at not getting the coffee as at my unshakable influence over him.

William stalked off to the fireplace. As he leaned over to reach the kettle, his lips curled back in a grimace at the smoke. He touched the handle of the kettle, then drew his hand back. He touched the handle repeatedly, till he found that it was not too hot to hold; then, carrying the kettle carefully and well away from his body, he went to the canteen. In the cup which was already a third filled with milk he placed two spoonfuls of coffee, then four spoonfuls of sugar. Finally he filled the cup to overflowing with scalding water from the kettle. He had my full attention, for the perfectly well-mannered way in which he had made himself a cup of coffee left me speechless.

It was a tin cup and too hot to be lifted, so William crouched over it, making extraordinary faces well before his lips were near enough to suck up any of the coffee. "It's hot, William," I said. "Be careful, it's hot." He glanced at me, his lips still curling away from the heat of the cup. Several times his lips came within millimeters of the hot liquid, but always he drew back before he actually touched it. He was impatient to drink the coffee, but realizing it was still too hot, he spooned some out, bringing the brimming teaspoon to his mouth. Finally he took a quick sip. Though the liquid had cooled considerably, it was still hot enough to make him jerk his head backward and drop the spoon. I expected him to empty the cup in disgust, but he didn't. He looked around, picked up several marble-sized stones and dropped them into the coffee. Surely, I thought, he can't know that by dropping cold stones into the cup he's going to cool the coffee? If he realizes that, what else does he know? It was a

William at work with a toothpick

disquieting thought. Could I have known William intimately all this time yet underestimated him so much?

He placed the spoon in the coffee and stirred it, then tried to sip from the cup again. The temperature just above the surface of his drink told him it was still too hot. He went to the jerry can, took a bulging mouthful of cold water, returned to his cup and spat the water into the cup. It overflowed, and he quickly sipped some of the liquid as it ran over the brim. The coffee was now hot but bearable. He lifted the cup and walked carefully to a shrub just beyond the kitchen; there he leisurely drank his coffee till the cup was empty. I quickly put the sugar and coffee back in the canteen.

I was able partially to overcome the problems with William by using a small starter's pistol that I had been given. The pistol fired blanks, so couldn't in any way harm William, but he was terrified of the bang it made and the small flame which shot from the end of the barrel. In really dire situations I could take the pistol out of my pocket and show it to him and he would immediately behave better. I rarely fired the pistol, just the sight of it had the desired effect. It took William some time to realize that I would not dare to fire it in front of Tina. By then Tina and I were on better terms and she was so established in the valley that I no longer feared she would desert me.

Soon after we came to camp, the fire began to hold a great fascination for Pooh and William—for William especially. He was perfectly well aware of the dangers of it and never really burned himself. He singed his arm once or twice, and invariably had to suck hot fingers to begin with, but he learned quickly by experience. I never saw him blow on the embers to ignite them, but occasionally he arranged them so well that they ignited spontaneously. William quickly learned to fill the kettle and heat up his own water. Indeed he acquired a great taste for hot water, which was not altogether helpful for a chimp who was supposed to be undergoing rehabilitation. Pooh was similarly corrupted and later on in the year, when the early mornings became cold, he would often spend the first hours of the day lying in the warm cinders of the fire, the firewood arranged in a star-shaped "nest" around him.

WE HAD BEEN IN CAMP just over two months. Tina was still nervous if I followed too closely in the bush, while William and Pooh would tend to stop and wait for me if they knew I was out walking with them. So, while Tina was pink, I frequently let them go off into the valley alone. Pooh sometimes stayed with them all day, but I suspected that these were the days when they remained in the area around camp.

After one such day with William and Tina, Pooh chose to sleep in an old nest in the kenno tree just above the shack. William and Tina were nesting in the gully. I was thrilled to see Pooh actually settle in a nest voluntarily, without any persuasion or encouragement from me. My delight proved to be premature. Just before I went to bed I filled the wash basin with water and washed my face and feet. Pooh must have heard me, for he left his nest and came down. He looked sleepy, but the soap and water was too inviting. Pooh washed his face and used my towel, and when I tried to put him back into his nest, he refused. Finally he returned to sleep on the platform as usual.

When Pooh was alone with me he reverted to being very dependent and whimpered a great deal if I made him walk. When we

Stella and Pooh

walked through open places I let him ride on my back, but it was exhausting work for me scrambling about with thirty-five pounds of chimp sitting on my back, his arms round my neck. I felt sorry for Pooh. He seemed to have always been a lonely little chimp. Each evening when Tina and William arrived, he was so obviously pleased to see them—patting them both in his excitement. William was usually callously offhand with him, but Tina would respond to his excited greetings.

The night came when Tina and William did not reappear at all. Pooh was reluctant to go to bed and came down to sit with me by the fire three times before he finally settled down to sleep.

The following morning I heard branches moving in the trees behind the shack. I leaped out of bed, and Pooh came hurrying down the ladder from the platform. We met at the edge of the gully and saw William and Tina walking toward us. William looked exhausted, his belly very flat, and when I opened the trailer to get out some food, instead of racing over and food-grunting, he merely managed a grin as I passed him his plate. I squatted beside him and placed my hand on his broadening back. "Tina wearing you out, Will?" I asked. He looked at me wearily, then put his arm round my shoulder and patted me with an open hand in the same way I pat the chimps when we share a few relaxed moments together or when I'm comforting them. It was such brief moments of complete communication with William that made up for all our squabbles and differences.

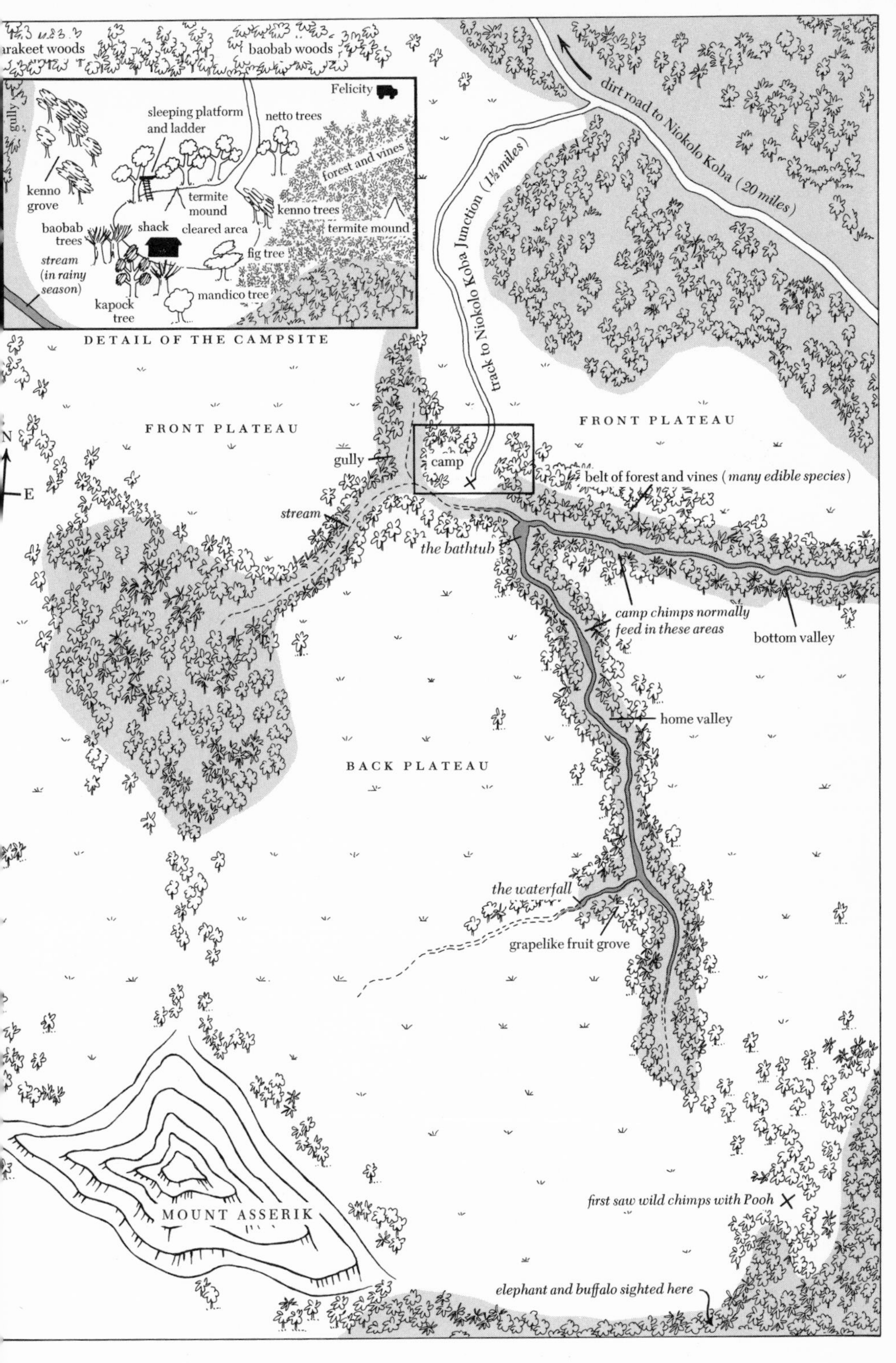

THE CAMP AT MOUNT ASSERIK

19. THE WILD CHIMPANZEES

After a night of rain, the chimps always spent the first part of their morning resting. After one particularly wet night I was pleased to see Pooh make his own nest and curl up in it. Tina made a quick nest and lay down too. William sprawled out on a branch. A little later I heard Pooh repeatedly slapping the branches around him. At first I couldn't understand what was happening. Pooh and William seemed to be displaying and shaking branches at a hole in the trunk of a large baobab that grew near camp. Tina was looking interested but not taking part in the displays. Suddenly I saw Pooh begin to slap his head, then both he and William half climbed, half fell out of their trees and began to scream and continually hit at themselves. They hurried toward me, and I realized that they were being pursued by angry bees. William shot straight past me, but Pooh leaped into my arms. Almost immediately I got stung on the shoulder, then on the neck. A continuous buzzing was going on in my hair as I raced for the shack. While unwiring the door, I got stung three more times. Few things induce panic so quickly as being surrounded and continually jabbed by irate African bees.

Finally the door was open and I dove under the mosquito net with Pooh. Several bees, either dead or dying, were still clinging to his fur and he continued to swat at them frantically. Several more bombarded the mosquito net in an heroic attempt to get at me. When I was sure we were safe, I began to inspect my stings. Pooh kept scratching his own stings and rocking madly with the excitement he'd experienced. I picked out the stingers that he had not already scratched out, and he did the same for me, delicately grooming the bee stings out of the inflamed red patches on my skin.

Once the immediate danger and panic were over, Pooh relaxed

and began to roll around with the bedding and pillow. I sat back and contemplated the throbbing bee-punctured parts of my anatomy. An hour later we cautiously ventured out of the net. Julian told me to take a bath and wash my hair, for the smell of battle I retained would incite other bees to attack me. I didn't know whether this was true or not, but I was not about to take any chances. Stealthily I crept down to the stream, frightening Pooh with my obvious nervousness. We met William on the way, and he joined us. Tina was nowhere to be seen. William's left eye was already puffy from two stings on his left brow. Each little sound that vaguely resembled a bee made me freeze in my tracks, while Pooh leaped onto my back or clutched at my legs. I succeeded in making William jumpy as well.

When we got to the stream, I jumped into the water as quickly as possible and began to scrub myself. My neck was swollen and stiff and the stings still throbbed. I gave William and Pooh a scrap of soap each so that they too could get rid of the smell of bees—just in case.

Shortly after this Tina left camp for an entire week. William was distraught. He looked for her everywhere, but never dared to go far from camp entirely alone. He again became a nuisance around the camp, only, without Tina there, he found me a much less patient person than before. Pooh and William devised what they considered a hilariously funny tease: unscrewing the valve caps on the inner tubes of the Land Rover tires and, with a long fingernail, pressing the valves and releasing the air. If I chased them, they ran round and round the car, always keeping out of reach, and if I got too close it was simple for them to slip under the car to the other side. Being so close to the ground, they could also keep a good eye on my progress. If they realized that they had a few seconds' lead, I'd hear a loud hiss as another blast of air was released from the car's already sagging tires.

A long bamboo pole and a handful of small rocks helped my pursuit, for with this equipment I could reach the chimps under the car far more easily. Once the Land Rover was no longer between us, it was simple for me to chase them away. I finally had to wire all the valve caps onto the car and cover the whole thing with cellophane tape. I kept expecting them to find a solution to that, but they either grew bored with the game or were genuinely foxed by my preventative methods.

While Tina was away I took William and Pooh on long exploratory hikes. We found some exciting new areas and food supplies, and Pooh seemed to enjoy himself thoroughly. William, for the most part, treated the walks with bored indifference, following yards behind on the way out and walking yards ahead on the way back. Upon returning to camp he'd immediately go to the gully, pant-hoot and wait, but Tina would not appear. He refused to build nests and reverted to sleeping on the platform. What was worse, he would leave the platform on wet nights and sneak into Julian's tent. Twice Julian awoke in the morning to find William sprawled on the ground sheet beside him. On the second occasion, as he was being firmly shooed out, William became indignant, dodged Julian round the tent, then grabbed Julian's blanket and raced up to the platform with it. He sat huddled in the blanket until the rain stopped. In addition to my growing anxiety for Tina, I began to feel increasingly sorry for William.

On one of our walks we discovered a large patch of forest not far from camp. There were several baobab trees in the grove, and Pooh and William climbed to eat the flower buds, as Tina had taught them to do. While they ate I walked round the tree and found four very fresh chimp nests. Suddenly a branch cracked nearby. William stopped feeding and, with his coat erect, stared in the direction of the sound. Julian and I quickly hid in a nearby thicket and waited. We heard no further sounds. William relaxed and continued feeding. When he and Pooh descended we walked on, but William kept very close to us and seemed wary.

At midday we stopped in the shade beneath a tangle of Sabba Senegalesis vines that were full of ripe fruit. When William had eaten his fill he came and sat close to me. I was writing at the time and was surprised when William reached up and began to fondle my hair, letting it run through his thick fingers and twining strands of it gently round his hands. When I smiled at him he held my glance for a moment, then rolled over as if bored. I continued writing, still smiling at his sudden affection after so much delinquent independence in the recent weeks. Then he rolled back to face me again and slowly reached and took my hand. Still holding it loosely, he fell asleep. It had been some time since William and I had held hands, and my hand felt very small in his larger one. I realized how fast he was growing and how independent he had become. His sudden but genuine affection made me want to hug him, but I knew he would consider that embarrassing, so I squeezed his hand quickly to return the affection and said, "I love you too, Willie boy."

That evening I lost William on the way home. He had been feeding on Sabba Senegalesis fruit with Pooh, and suddenly vanished. Julian and I checked the immediate area, then decided that William must have gone home alone. Camp was still a good two miles away, and William's sudden desertion struck me as singularly odd. William disappearing for a day with Tina was one thing, but vanishing in the middle of a walk, alone in the bush somewhere with night approaching, was quite another. We returned to camp as quickly as possible, walking so rapidly that Pooh began to whimper. I walked into camp to find the place deserted.

Panic gripped my stomach. Julian and I retraced our steps, Pooh riding on my back. We walked to where we had last seen William, and I called him. Finally, forced by time, we had to return. As I trudged into camp, exhausted from the anxiety and from carrying Pooh on my back for two hours, I heard the beginning of a pant-hoot that broke into screaming as William ran into camp from the gully and embraced me. He had obviously been afraid by himself, and even more afraid at finding camp empty on his return, and was as relieved to see us as we were to see him. I gave the chimps their supper and a cup of milk each, and they went to bed.

That night there was another torrential downpour. The gully stream swelled to river proportions, and muddy water swirled ferociously around the rocks and fallen trees, carrying branches and debris with it as it pounded on. The crystal-clear trickling stream had been transformed into a raging waterway.

The next morning I gave Pooh and William as much breakfast as they could eat. Then both chimps lay out in the watery sunshine to catch up on the sleep that the rain had prevented. Julian and I set about clearing the camp of the mud and leaves the water had banked up against the sides of the shack and the tent.

Shortly after lunch there was a chorus of excited pant-hoots in the vicinity of the valley. A second, closer group answered, their calls and excited drumming on tree trunks clear and thrilling on the clean, rain-washed air. I dropped everything and ran to Pooh and William. William was standing, staring across the back plateau in the direction of the sound, his coat bristling. Pooh leaped into my arms. Both William and Pooh answered the second chorus of wild chimpanzee calls, and seemed to receive replies.

The valley seemed full of vocalizing wild chimps. I ran to the shack to grab my binoculars and camera. I asked Julian to say in camp, as I thought the wild chimps would be less afraid of one smallish human being than two. Pooh followed me eagerly, but William looked extremely reluctant. Each time the wild chimps pant-hooted he grinned at me nervously and half turned to go back.

I coaxed him on. The sodden ground made it easy to walk silently. We followed the calls for about two hours. The valley was now behind us, and we entered a sparse woodland.

Suddenly there was another burst of pant-hooting. This time I couldn't be mistaken—the wild chimps really were just ahead. I could hear branches cracking, and a juvenile screamed. I looked round to find William gone. The calls were so deep and vibrant and numerous that I felt my adrenaline begin to rise, and when I looked through the binoculars there was a slight tremor in my hands. Pooh climbed onto my back and threatened to whimper when I tried to put him down, so I let him ride. I wasn't sure whether William was ahead of me or had gone back.

Then I saw them. There seemed to be four groups of them, feeding. One was just ahead of me, another was in a gully, and two more groups called from the edge of a plateau farther along. I crept closer. I could see them feeding in the vines at the head of the gully. The vegetation was so thick that it was difficult to see them clearly. There was a terrific amount of calling and food-grunting all around me, but I had only glimpses of a face, a back or a long arm as it reached out to grasp a fruit.

I followed a line of shrubs and stunted trees onto the plateau, moving as close as cover would let me to one of the feeding groups I had spotted. I sat about thirty-five or forty yards away from them, Pooh at my side. He was intensely alert to their sounds and movements, but he remained passive, showing no inclination to answer or to approach the wild chimps. I glanced to my right, and a sudden movement caught my attention. The ground vegetation made it difficult to see properly, but—not far away—there were at least three adult chimps gathered around a termite mound. I looked through the binoculars, but that didn't help. I was sure that the chimps were fishing for termites just as I had seen the wild chimps at Gombe do. I didn't dare move for fear of frightening them all.

For over ten minutes I sat, enjoying to the full being so close to the wild chimps of Mount Asserik. One adult chimp walked out of the gully which adjoined the plateau, and I could see him quite clearly. He climbed into a small cherry tree, sat on a branch and looked round. As his gaze fell on Pooh and me, he started visibly. He glared for a second or two, every hair on his body erect, then made a series of loud, aggressive whaa barks in my direction.

Suddenly every chimp hidden in the vines took up the cry. I wondered whether I'd been stupid to creep so close unobserved—whether instead of running away, the wild chimps would attack us. The noise was deafening. Pooh clutched me tightly. I got up as slowly and as casually as I could, Pooh climbed onto my back and we walked away from the chimps and sat down again, this time in the open. I hoped to indicate that I meant no harm. When I next looked at the vines, there wasn't a chimp in sight, but a few more whaas were directed at me from the thicket. Then all was quiet. A moment ago the area had vibrated with the calls of feeding chimps—now the silence was eerie.

Pooh and I sat for half an hour until I thought the chimps had left the area. When I stood up Pooh again climbed onto my back. I began to walk slowly toward the forest, then stopped. There, barely twenty yards away, on the forest's edge, six adult chimps stood, silently watching me. They were enormous. I had always considered Tina large, but these characters looked incredibly tough and capable. As soon as I stopped, they turned, one by one, and walked quickly into the woods. Several stood and looked back for a moment, then disappeared. I thought one of them must be quite old, for the hair on its back was sparse and reddish, instead of the usual black.

I decided to keep walking in the open area so that any other chimps watching would see me clearly. I didn't want to give the impression that I was stalking them or doing anything suspicious. Halfway across the plateau, I heard them pant-hoot again. They hadn't moved far away.

I decided to return to camp. Tempting as it was to stay, it was probably best to expose ourselves to the wild chimps for short periods to begin with. I did not want to frighten them. I knew there would be more opportunities to watch them.

I was sorry that William had left us at the last moment, and the more I thought about his nervous reaction to the first wild chimp calls, the more convinced I became that he had returned to camp. Perhaps he and Tina had already met wild chimps while they were out alone and had not received a friendly reception.

When we returned William came loping out of camp to meet us, and gave us both a terrific welcome. It had been an exhilarating day for me, and I spent the entire supper telling Julian about the encounter in detail.

For three consecutive days we followed the wild chimps' calls, returning to camp each evening. It was a strenuous period for William and Pooh, for each day the chimps moved farther round the mountain, and farther away from camp. For the last two days William and Pooh walked almost ten miles each day.

Walking with our two chimps was slow, so we were never in close proximity to the wild chimps for more than a few hours and I was never able to get as near to them as I had been on the first day. William continued to be uneasy and timid when he realized that we were approaching the excited calling of the wild chimps. When the time came to head back to camp, he led the way willingly. By the end of the third day the wild chimps had moved so far away that we couldn't catch up with them before it was time to return to camp.

When we reached camp that evening a wonderful surprise awaited us. Tina was sitting in one of the trees behind the shack. She panted at me affectionately, and when William heard me greeting her he raced into camp. Rising on two legs, he strode toward Tina's tree, his coat erect. She descended, panting hoarsely, and they embraced, squeaking and grinning their excitement.

20. A NEAR MISS

TWENTY-FOUR HOURS AFTER TINA RETURNED, I was almost wishing that she would disappear again. William switched instantly from being a peace-loving, if somewhat irritating, young chimp, into his more typical role of tyrant. For the first day after Tina returned, he appeared drunk. The morning began with William trying to overturn the frame tent, then leaping up and down on its roof till it threatened to tear. I managed to catch him, slapped him hard and told him to stop it. Just to show how remorseful he was, he punched me back, then somersaulted halfway across the yard. Suddenly he leaped up, charged the kitchen table and, as he rushed past it, skillfully grabbed the tarpaulin table cover and hauled the whole thing off. He cantered on a few more yards, then stopped, panting. He sat in the middle of the tarpaulin, looked at me and rolled backward. Laughing heartily to himself, he began to wind himself up in the tarpaulin.

The tarpaulin was essential to keep the firewood beneath the table and the canteen dry, and there was no question of letting William ruin it. I strode over to him, looking so purposeful that normally he would have at least sat up to take notice. Instead he just carried on laughing and rolling about. Tina was on the edge of the gully when I reached him. I hissed at him to behave. There was not the slightest reaction. I began to think that he really was drunk or high on something. I held his shoulder and shook him as hard as I dared in front of Tina. "William, what's the matter with you!" He looked at me lazily, shrugged my hand away and somersaulted backward, still keeping hold of a corner of the tarpaulin. I gathered up the rest and pulled. William pulled back and picked up a rock in his other hand. I felt confused. There was nothing aggressive in his relaxed, floppy posturing, yet I had the feeling that if I continued to pull, William was going to sling his rock at me. Ever since we had left Abuko I had been waiting for William to choose his moment for a real fight to re-establish which of us was dominant. Until then I had always held that position. Now William, at eight years old, found he could no longer automatically accept my position in the camp hierarchy.

I was not afraid of William—I had known him too long and too well—but I didn't want a showdown in the middle of the yard, right under Tina's nose, because I *was* afraid of her. She was big and powerful, with a formidable pair of canines, and, more important, she was an unknown quantity. Slowly I took the starter's pistol out of my pocket. The ruse worked. William dropped his rock and laboriously rolled away from me, letting go of the tarpaulin.

Julian and I spent the next half hour wiring the tarpaulin onto the table. As we finished, William walked to the shack. The door was unlocked. He opened it, then slammed it back. I heard an ominous crack as some of the wood splintered around the hinges. "William, stop it or I'll thump you!" I shouted.

Tina swayed. William sat—his coat semi-erect—and glared at me. I strode over to him, anxious to avoid deliberate provocation while Tina was around, yet determined not to lose face. I closed the door and locked it. William got up and punched my foot. By this time it was difficult not to react. I found his insolent taunting unbearable. When he swung round, grabbed my ankle and neatly flipped me over, I could stand no more. The next time he approached, I kicked out hard and caught him in the thigh. He screamed, his arrogance flown with the wind. I walked over to him and cupped his chin in my hand. "Now then, will you stop bugging me, you brute!" He listened, then pulled away sulkily and walked toward the edge of the gully. Tina was glaring at me, but fortunately the shack had hidden most of the incident from her. I had maintained my position of authority. I knew that if William ever succeeded in frightening or intimidating me, my days in camp would be impossible. Tina, William and Pooh stayed out of camp for the remainder of the day. I watched them for a few hours, then returned to camp. When William came back that evening he greeted me affectionately, and though he continued to be boisterous

William tries to devise a way to break into the shack.

he had lost the insolent, provoking manner that he had shown at the beginning of the day.

Around this time I wrote to Michael Brambell, the curator of mammals at the London Zoo. He was still very interested in sending out Yula and Cameron to see if their rehabilitation in the wild would be possible. I remembered vividly the two pale faces looking at me from behind the bars of a cage: Yula's quiet, intense expression and Cameron's excited gestures as he played with Dr. Brambell and his keeper. I imagined them here with Willie and Pooh and Tina—in the vines behind the shack, walking in the valley and feeding in the tabbo on the edge of the gully. Willie and Pooh might just as well have been born in captivity for all either of them remembered of their natural mothers and regions, but they had grown up in a richer environment than Yula and Cameron had. Abuko had been an ideal training ground for their present lives. I wondered how Yula and Cameron would react—whether they would ever catch up with the experience needed to live such totally new lives. I wasn't sure, but William and Pooh had settled into camp life so easily that I wanted to give Yula and Cameron a chance. They would have the advantage of being able to learn from three other chimps. Also, I had gained more experience, and would be better able to guide them in the right directions. I wrote Dr. Brambell a long letter, saying that the rehabilitation of William and Pooh seemed to be progressing well, and that I was eager to give Yula and Cameron the same opportunity.

For two days Pooh had stayed constantly with Tina and William. There was plenty of ripe fruit in the gully, so Tina was keeping close enough to camp for William and Pooh to remain with her all day. I watched them through binoculars from the gully edge to interfere as little as possible. On the evening of the second day, Pooh sat at the top of a high tree and studiously watched Tina make a nest. Minutes later he descended into the thicker foliage, and I heard branches being broken. Pooh was either making his own nest or renovating an old nest of Tina's or William's. I could not see him from where I sat and dared not approach in case he should notice me and return to camp. William had seen Tina and Pooh make their nests in the gully, but when I suggested it was bedtime, William went to the platform. There were no leaves there, but he lay on it anyway.

While we were having supper, William appeared in the kitchen. He had already eaten and when he realized that there was nothing more forthcoming he sat on the edge of the gully. Finally he disappeared into the shadowy vegetation. It was with a sense of achievement that, minutes later, I heard him making a nest. For the first time all three chimps were using nests in the gully outside the camp area.

William woke me early the following morning by pulling on the wire mesh that covered the door of the shack. I got up and gave him and Pooh their milk. Pooh, who hadn't been into camp since the previous afternoon, surprised me by rushing back to Tina the moment he had finished his milk. She had no swelling, so William was in no hurry. He sat around watching me wash, stole my toothbrush the moment I put it down and ran off. He sat for several minutes on the edge of camp, cleaning his teeth in exactly the same way as I'd cleaned mine. He then sucked the toothpaste out of the brush, left the brush on top of a rock and went to the gully to join Pooh and Tina.

I took the binoculars and went to find the chimps. The gully was deserted. I walked to the main valley stream and followed it down about a mile, but still found no trace of them. Perhaps, after two days with Tina, Pooh had gained the extra confidence he needed to follow her farther afield. I backtracked through camp and went to look a little farther up the valley. Finally, at about lunchtime, I returned to camp to wait.

Suddenly the terrified screaming of a chimp reached me faintly across the plateau. I froze and listened with every nerve of my body. The chimp screaming was Pooh. I could recognize his voice among thousands. I ran out onto the plateau. Julian followed. In the distance I saw William loping toward me, leaving the tree line and running out onto the rocky edges of the plateau where there was scarcely any cover. Then he vanished into the grass. When I reached him, he hugged me. I left him with Julian and ran on to where Tina was swaying in an isolated wild cherry tree. I looked frantically into its branches for Pooh—he was nowhere to be seen. Where was Pooh? Why had he screamed? Farther up the gentle slope, well in the tree line, baboons began barking.

My stomach screwed up into a tight ball and my chest felt as if it had turned to stone. I thought I knew what had happened. Pooh had seen the baboons, and had run out onto the plateau to chase them, so engrossed in his daring that he was heedless of the open terrain. A lion that we had heard roaring in this very spot the night before was probably lying in the tree line. Pooh would have been easy prey.

Tina swung out of the tree and ran up the slope, then stopped, looked back at me and hurried on. In my distress I took her behavior to mean that I should follow her. She hurried up the slope, then paused on the edge of an enormous basin that dropped down on the other side. She plunged over the steep edge of the basin and disappeared at a run into the long grass. I hurried after her but quickly lost her. I tore through the grass, too worried to be cautious. William was running after me. We found trails, we followed waterways, we saw a set of chimp tracks, but they were large—perhaps

Tina's, certainly not Pooh's. I called and called Pooh's name. I was hoarse, my lungs ached from the chase and it hurt me to shout, but the worst pain of all was the torture which was steadily growing, the agony of believing that I might never find him, that Pooh was dead.

Every muscle trembled and my clothes were soaked with perspiration as I climbed back up the slope to the plateau. The one hope left was that Julian might have found Pooh while I was searching in the dip. I hurried to the edge of the plateau and saw him walking dejectedly by himself back to camp. We had been searching for three hours. Hopelessly, I stumbled to the limited but welcoming shade of the cherry tree where I had first seen Tina. William was close behind but overtook me as we neared the tree's limited but welcoming shade. He sat down, draping one arm across a large boulder, and stared across the plateau. Tiny lights starred his face as the sun was reflected in the minute drops of perspiration there. He looked hot and tired, but relaxed.

I searched the ground for signs of blood. In my aching, empty head I heard again Pooh's voice screaming fear, cutting the sun-drenched quietness of the plateau, reaching me too late. I embraced the rough branch of the solitary cherry tree and let it support me. Then I prayed with an intensity and fierceness I didn't know I possessed. As if in answer, a lion roared, then coughed his call to its mournful end. William touched my leg. He was holding out a dry leaf to me. When I didn't immediately accept it, he tossed it toward my feet, but the leaf, being light, settled closer to his own. He picked it up again and shredded it.

William, of course. William. Of all the hopes, the plans, the achievements, only William and I survived. I had failed. But Pooh was to be my last failure. I would take William home to those who could look after him. Yula and Cameron would spend their lives safely in a cage. I only wanted to die, to stop feeling the guilt, the unbearable burden of the responsibility for such precious lives.

William could perhaps sense but not comprehend my uncontrollable sobbing. He handed me small stones, a twig, and finally patted my back as I had often patted his and Pooh's. Then he moved away, striding out for camp. I struggled to my feet and followed him. The landscape was blurred and distorted. I stumbled on a rock and fell.

William stopped and watched me get to my feet again, then strode on.

Julian was standing in front of camp, waving to me. I wiped my face with the bottom of my shirt and stared. He was holding something. My heart began to hammer so hard that I could feel its vibrations in my head. I lifted the binoculars—hope making me feel sick. Pooh was sitting on Julian's hip, looking out at William and me as we crossed the plateau. I ran, fell down and ran again. As I ap-

A hug from Pooh

proached, Julian put Pooh down and he ran toward me. It seemed a miracle to be able to hold him, to see his wrinkled, gnomelike face again. I rejoiced a thousand times, with as much passion as I had begged for his return. I examined him minutely for injuries. Apart from a fresh graze on his thigh, he was perfectly all right.

I held him tightly all the way to camp, and he occupied himself by licking the salty tears off my face and neck. When we got to camp, I promptly vomited. Julian lit the fire and made us all large mugs of tea. William asked for a second mug, and his cup was refilled. Julian told me that he had searched the lower regions of the stream again, and as he was returning to camp had heard a whimper behind him. Pooh was hurrying to catch up. To this day I do not know why Pooh screamed or where he was during the time we spent searching.

By evening Tina was back in camp. She was exceptionally affectionate, panting and holding out her hand whenever she passed me. Pooh was quiet for several hours and remained close to me, but he had recovered when the time came to nest. He climbed into an old nest of Tina's just behind the shack. After pulling in fresh foliage and fastidiously arranging it, he lay down to sleep. William disappointed me by climbing up onto the bare platform and lying out there.

That evening I felt exhausted, and the horror of believing Pooh dead stayed with me. It was the same feeling as when one wakes up in the night after a terrible nightmare. The relief of knowing that it was all a dream doesn't immediately dispel the fear and anguish experienced so vividly.

21. PASTIMES AND PRACTICES

NEXT DAY I WANTED ONLY to stay in camp, but the day after that I decided that, whether or not I felt so inclined, I must take the chimps exploring again. I decided to go to a large basin at the foot of Mount Asserik. I remembered it as being full of fig trees, and they were probably all fruiting. Julian and I left camp at seven o'clock with a packed lunch.

It was a long walk to the dip. We followed the valley to its head, stopping periodically to allow William and Pooh to feed. Then we crossed the woodland area where I had seen the wild chimps, and a mile or so farther on came to the lip of the dip. As I had expected, the grass was high, so we walked round to where a gully entered the dip. A rainy-season stream led into the dip, and walking along the edge of this natural drain proved fairly easy. We were just coming onto the floor of the basin, heading for the grove of fig trees, when a rumbling sound made me swing round to my right. A young elephant was calmly grazing about thirty yards away from us.

On closer investigation I found that there were three elephants —two young ones and an adult with long, straight tusks. I took some photographs. William and Pooh crept around with us, intensely interested. They seemed to be observing me and imitating my reactions. We sat silently watching the elephants for about ten minutes, then they began to walk toward us, grazing as they went. We crept away. Pooh climbed onto my back, and the elephants knew nothing of our presence.

William led us quickly to a large baobab tree and he and Pooh climbed it and fed on the flowers, constantly looking around and obviously ill at ease. Suddenly I glanced up. An enormous elephant was approaching us, having just left the thickest part of the fig grove. Julian and I slipped away from the baobab. William scrambled down and crept silently to us. Pooh seemed hypnotized by the enormous bulk ambling directly toward us. "Pooh, come here!" I hissed as loudly as I dared. Pooh didn't move. I crept on, hoping Pooh would follow. The opposite happened: the farther away I moved from the baobab, the less courage Pooh could find to leave it. He slid to the lowest branch and kept glancing from the elephant to me, judging whether there was time to run for it.

The elephant strolled closer, then reached up with his trunk, broke off a kenno branch and began to eat the young leaves. Pooh's lips drew back in a silent grimace of fear. William began to whimper, but stopped the instant I turned to him with my forefinger held up to my lips. I told Julian I was going for Pooh. If the elephant found out we were so close, he would probably run away, but if he charged, Julian was to run. Julian said I was not to worry on that score, he didn't intend to face a frightened elephant alone. I crept back to the baobab, keeping the tree between me and the chewing elephant. Every rustle in the grass seemed amplified, and I was sure that at any second I would be discovered.

Pooh watched me approach. I got to within about ten yards of the baobab tree and held out my arms. I hoped frantically that Pooh would not lose his nerve and leap out of the baobab tree with a thud. He hesitated for a brief second, shot a glance at the elephant, then quickly but silently slid down the smooth trunk of the tree and ran to me. As we made contact, he gave a high-pitched squeak of relief, then clung to me tightly. We both looked back at the elephant. His audible chewing had stopped. He was no longer slowly flapping his ears against his neck as he enjoyed his meal. He was standing like a statue, listening, the tip of his trunk exploring the air for alien scents or hints of danger. I froze, certain that at any second he would smell me or hear the thudding of my heart.

He stood suspiciously for what seemed an eternity, then swung around and walked away quickly, stopping at intervals to listen and smell. I waited till I considered him a safe distance away, then crept back to Julian. I was pleased with the chimps for reacting so well. As soon as I reached Julian, William began to lead us out of the dip. He was frightened, constantly standing upright to look out over the grass and hurrying along as quickly as he could without actually running. Pooh clung to my back, his tight grip telling me that he too was far from relaxed.

We had even more reason to be disturbed by our next encounter. We had not gone far back down the gully before we found buffalo tracks and dung that was still almost steaming. The tracks crossed the gully rather than following it, and we presumed that the buffalo had continued into the long grass.

We set off again, keeping to the edge of the water. The smell of the buffalo seemed to stay with us, giving the impression of walking through a farmyard rather than a national park. William, slightly braver now, was again leading the way. The tremendous crashing of vegetation, William's scream and Pooh's gripping my neck seemed to happen all at once. Julian got into a tree almost as fast as William. I stood rooted to the spot, Pooh clinging desperately to my back. The vegetation all around seemed to have burst into violent life. Then, about ten yards ahead of me, I saw the fattest, roundest buffalo I'd ever seen. His head was lifted and moist black eyes stared stonily down his nose in my direction.

A few yards behind me a vine hung from a high branch. I grabbed it, but with Pooh's extra weight on my back it took every ounce of my strength to get even halfway up. Pooh refused to leave me and climb ahead alone. The day had been one long, nerve-wracking experience, and Pooh was sticking with me. The bush all around was still thundering with stampeding buffalo. My arms were getting weaker. If Pooh didn't get off my back, I knew I was going to slide down the vine again. There was little to hold on to.

I glanced down to see where the buffalo was, and began to slip. It was difficult to slow my rate of descent, and when I reached the ground my thighs and hands burned, my fingers ached and refused to open. I made a mad dash for Julian's tree, and he hauled me up onto the lowest branch. Then he persuaded Pooh, who was still clinging desperately to me, to come to him. After attempting to catch my breath, I tried to climb higher. I was shaking all over and my knees felt like water. I had to control a terrible urge to laugh. Julian told me that as I ran for the vine, the buffalo snorted, then spun round and galloped after the rest of the herd. We could still hear their progress, but the sounds were fading fast. I could hardly believe it. All the times we had been out and seen nothing except antelopes, baboons and the odd warthog, and then in one day we almost touched noses with elephants and buffalo. We waited in the tree for half an hour, talking normally so that if any buffalo still remained in the vicinity we would not surprise them.

Things are always said to happen in threes. The next morning a worried whimper from William suddenly attracted my attention as I rested peacefully in the gully. Through the binoculars I could see him looking into a hole in the rocks. Every hair on his body stood erect, and he frequently glanced back at me. I hurried to where he sat. As I drew near he picked up a large rock and flung it with considerable force at the ground a yard or so ahead of him, then jumped back onto a boulder. On account of the grass and low vegetation, I could not immediately see what was troubling him so much. The closer I got, the more confident and aggressive William became. He hurled another rock underarm, then another; then he picked up a heavy stick of dead wood and used it as a club. On the second blow he let go of the branch and leaped backward again. Pooh, who was right behind me, saw the snake before I did and rushed into a nearby tree, whaaing loudly. Seconds later I saw a beautiful royal python, about four feet long and in superb condition, frenziedly trying to climb a sheer piece of rock to return to its hole. Apparently one of William's rocks, or the club, had hit the snake, for there was a graze about a foot from the end of its tail, but otherwise it was unhurt.

I knew the python was harmless, but for William and Pooh's sakes I feigned exaggerated fear, grabbed Pooh and backed away. William retreated initially, but Pooh kept making raucous, excited whaas and William rushed in again with another branch. I whimpered and continued to back away, calling to William urgently. He took no notice, he was too involved in his attack on the python. I picked up a small piece of wood and threw it at William, making the low alarm hoot at the same time; then I ran furtively up the slope toward camp. William stopped his show of bravado and hurried after us.

WE HAD BEEN IN CAMP for 6 months now. The rainy season had begun and, with the rains, the grass had shot up. Even on the rocky plateau it reached waist-high. We kept an area of ground in front of camp free of grass, but the moment I left this bare patch my trousers and canvas boots were instantly soaked. Around midday, if it didn't rain, I usually dried out; my boots, though, were frequently damp right through the day, and I got a severe case of foot rot. The soles of my feet became pitted and then raw, till it became painful to move. When Julian saw me tenderly inspecting my feet, he asked why I didn't wear the special water boots. I realized he meant wellingtons. I had thought they would be so hot that I would get foot rot from perspiring, but I decided to try. They turned out to be ideal. They were hot, it was true, but much easier to slip off and on. My feet stayed comparatively dry, and the foot rot quickly healed.

It had been over a month since anyone had visited us, and food supplies were getting very low. There was still plenty of rice, but virtually nothing left to eat with it except some dried fish. I began listening for the sound of a car approaching and hoped Claude would come soon. The next few days passed quietly. William was so besotted by Tina at this time that he came into camp in the evenings for no more than several minutes. Pooh spent only a few hours with Tina and William each day, because William was so possessive of Tina that he made Pooh feel unwelcome. Tina attacked William once for chasing Pooh and made William scream with frustrated rage. He was not confident enough to retaliate, being smaller than Tina, and finally threw himself into a bush in a terrible temper and rolled about until his screams choked him. Then he sat up, calm once more, and whimpered to Tina. She ignored him at first, but he followed her so persistently that she eventually turned and presented to him and all was peaceful again.

During the afternoon Julian went out for a walk alone. He was late returning and I decided to make a start on an early supper. I took some rice and some of the little dried fish that remained and had just finished cleaning the rice when Julian came back. In one hand he held a clump of ripe kabba fruits, in the other a basket made of huge tabbo leaves pinned together with twigs. He opened his homemade shopping bag and took out a perfect mushroom. It was in the last stages of opening from its button shape and looked firm and fresh. The upper side was a powdery white with delicate brown speckles, the underside a healthy dark brown.

Julian's basket was full of equally fine-looking specimens. I could feel my mouth watering. How could anything that seemed so deliciously edible be poisonous, I silently reasoned, but finally I said to Julian:

"How do you know they are good to eat?"

"We eat them every rainy season at home," he replied. "My mother picks so many that sometimes she has to dry them and keep them for later."

"Are you sure these are the same ones, Julian? Mushrooms can be very bad for you, you know."

He gave me a pitying glance and assured me that they were not only edible but tasted very good—just like sweet meat. I'd been craving steak for weeks, so I stopped talking and watched him cook. While he cooked Julian told me all he knew about mushrooms. One of his facts was that all mushrooms that grew on rotting kapok trees were edible. This statement sounded so general that I almost let caution triumph again, but the smell was now so good that I could hardly keep from drooling. Julian served the mushrooms with rice and fried kabba fruit. I let him start and watched carefully. He chewed, swallowed and made appreciative noises. Still, it might be an hour or two before anything happened. In the end I compromised by taking a sample mouthful and reserving judgment till the next day.

Julian was up before me the following morning, clearly healthy and whistling to himself as he made the fire. While William and Pooh ate their rice and milk breakfast, I savored a plateful of the finest mushrooms I had ever tasted. From then on I became as avid a collector as Julian. There were four different types of mushrooms that he approved of. The rest he assured me were not poisonous, they just did not taste good. I quickly learned which trees fostered our favorite mushrooms. Often we got to them too early, when the mushrooms were still tiny buds; sometimes we were too late and they were already rotting; but we found them in ideal condition often enough to keep us enthusiastic.

Occasional mushrooms helped, but our evening meal was still usually a rather miserable affair of dried fish and boiled rice. And since the sugar had been finished and milk was strictly rationed, there wasn't even a good cup of tea to liven it up. I was beginning to get seriously worried that we were going to have to reverse the rules and let the chimps rehabilitate us.

I remember one evening vividly. It was beautifully clear after all the rain the night before. The gully stream had calmed during the day till it merely gurgled enthusiastically instead of bellowing. The flying termites were out in force, making it difficult to sit too near a light. I was dozing off when Pooh, who was sleeping in a nest above the shack, began to pant-hoot. Tina and William joined in. I got up and walked outside with the flashlight. Pooh was immediately at my side. In the distance I could hear the sound of a vehicle. It gradually came closer, till I was able to distinguish headlights bobbing along in the grass on the plateau. A Land Rover pulled into camp. It was impossible to tell what color it was, not because of the limited light, but because it was literally caked in mud. An equally muddy Claude stepped out. It had taken him all day to travel the last sixteen miles to camp.

When Claude went off again next day our little kerosene-powered fridge was bulging with fresh vegetables, meat, cheeses, eggs and—the greatest luxury—a pound of butter. Best of all, Claude had brought with him Réné, who had helped us at the time of the first release and was now to stay and keep us company.

SEVERAL NIGHTS after Claude had departed, I awoke suddenly. There was a brilliant moon. Everything was quiet and calm, but I was sure something had brought me out of my deep sleep. I listened for a moment, then began to doze off. The sound of the tarpaulin being lifted carefully from the roof made me start to consciousness again. Something was trying to creep into the shack. I lay frozen, my eyes fixed on the dark corner where the rustling sounds were coming from. I reached out slowly and picked up my flashlight, pointed it at the sounds and waited. The rustling grew louder and more determined. The tarpaulin was now being pulled upward with some force. I flicked on the light—the beam illuminated a large, hairy black arm and William's startled face.

I got out of bed, unlocked the shack door and stormed out.

William was already loping away into the gully. I followed him to his nest, giving him stern orders to go to bed. I watched him climb through the dappled moonlit shadows and into his nest. Then I returned to the shack. I felt sure he would not be back before morning, and decided to wait until then to rewire the tarpaulin to the roof.

Some time later I awoke again. It was pouring outside, and a fine spray was coming through the mesh on the windows and settling on my bed and my face. To roll down the tarpaulin over the windows I'd have had to go out and get soaked, so I decided I'd move the bed nearer the wall, where I'd be less likely to get wet.

I reached for the flashlight. It was not there. Figuring that I had probably left it by the door after showing William to his nest, I sat up and felt around on the floor with my feet for a pair of flipflops. I had to control a squeal of horror—my toes sank into something cold and soft. I whipped my legs back onto the bed; my foot was covered in a sort of grease. I fumbled around again and found a box of matches. The flame flickered just long enough for me to glimpse the wreckage in the room.

My God, I thought, there must have been a strong wind with this rain! The second match illuminated what I had stepped in—half a pound of butter. Now, *no* wind was strong enough to blow the precious butter out of the fridge. I suddenly became very suspicious. By the fifth match I had found the flashlight. It was in pieces by the bed, but the batteries were lying beside it. I screwed the flashlight together again, and slipped in the batteries. Everything still worked. I shone the light round the shack slowly. There was something very familiar about the lay of the wreckage—this was no wind; this was William.

I continued shining the light around, assessing the damage. My camera? No, that was still in the tin on the canteen. The fridge was open, and everything that had been in it now lay on the floor—jam, milk, the remains of Claude's pâté, butter and a small piece of meat. There were also a couple of orange peels and the remains of half a cabbage. Farther away from the fridge, a wellington boot and some clothes, binoculars unscrewed into pieces, notebooks, a packet of chewed felt-tip pens, a lighter, an open penknife, a tube of toothpaste, insect repellent. An empty bottle of wine, more clothes, another jar of jam and several long loaves of french bread. A foot. I flicked the light back—yes, a foot. William had made a nest of clothes under the table and was in a drunken sleep. Claude had left a bottle of *vin ordinaire* which was sealed with an aluminum cap like a Coke bottle. William had obviously opened it with his teeth and drunk the whole thing. I felt so angry with him for causing such chaos that I would gladly have thrown him out in the rain. I was afraid, though, that after a liter of wine he would be sick, so I let him sleep on. He didn't stir as I tidied up as much of the mess as I could in the dark.

I awoke early the following morning, before William. It had stopped raining. I walked over to him and shook his shoulder. He stirred and rolled over. I shook him again. He sat up, blinking, and looked around, which relieved me. Then he saw the bread, and reached for half a loaf. That decided me. I unwired the door and ushered him out.

During the next few weeks Tina spent more time away from camp than she did in it. William seemed desperately bored when she was away, and though he was deliriously excited to see her each time she returned, he only stuck close to her if she had a swelling. Otherwise he would let her wander away and follow me instead. The days Tina appeared I tried to stay in camp so that William and Pooh would remain with her. This seemed to work less as time progressed. If I remained in camp, Pooh and William would spend some time in the gully with Tina, but when she left they would return to camp and hang around getting bored. Sometimes Tina would stay with us all day, but more often than not she would follow us for a few hours, then disappear. We got to know the Mount Asserik area well. Each day either Réné or Julian would walk with me while the other one did the camp duties. We left camp early, taking a flask of coffee and a light lunch, and usually did not get back before six o'clock in the evening.

Sometimes while we were out the chimps got thorns in their feet. They usually tried to pull them out by themselves, but if the thorn was a particularly stubborn one, the chimps would come to me and show me where it was embedded. Though William and I occasionally fought, he had implicit faith in me when he wanted help and would sit for minutes on end while I picked deep thorns out of his skin. Occasionally I could not help hurting him in the process, but he would only wince, suck his foot or hand and give it back to me. I had a Swiss army knife that contained a minute pair of tweezers, and with this and a safety pin I was almost always successful in extracting the thorn.

William was surprisingly ingenious at treating his own minor ailments. When he got an ear infection, he would frequently clean his ears with twigs and bird feathers, twisting them back and forth between the sides of his forefinger and thumb in much the same way as people use cotton swabs. After a bout of sneezing or when something irritated his nose, he would push tiny pieces of grass stems up his nostrils, where he would leave them till he sneezed or snorted them out again. He used twigs for picking his teeth, a habit he probably learned from Julian, who often whittled a twig to size and used it as a toothpick. Pooh was equally inventive in his own way, but concentrated on toys and means of amusing himself. He loved looking through the binoculars and often reached for them when I was using them. I always held the binoculars while he looked, as I didn't trust him to handle them with the care they required. Invariably I got tired of holding them before Pooh got tired of looking, and when I put them away Pooh would make his own play binoculars by placing a small pebble in each eye socket and screwing up his face to hold them there.

After the construction of the shack, Pooh became an enthusiastic carpenter. He seemed to get most pleasure out of hammering. In camp he used a length of bamboo to hammer down the heads of the nails which held the shack together. He hammered pieces of wire into the ground. He made a drum from his tin food bowl by hammering it. He hammered on the aluminum bowls. The more noise his hammering made, the better he liked it. Later he learned to put his hammering to practical use and hammered open fruits too hard for his milk teeth to cope with.

THERE WAS AN ENORMOUS baobab tree close to camp. It had a massive, smooth girth, and the first branch occurred about ten feet up the trunk. William tried persistently to climb the trunk, but it was too wide and smooth for him to scale. There was no convenient neighboring tree which might have afforded a crossing through its branches, and it seemed that the few fruit still dangling temptingly in the higher branches were going to stay there. But William did not give up so easily. One day, having exhausted himself in attempts to shinny up the trunk, he sat on the ground and recovered his breath. Then he got up, walked purposefully to a small fallen tree and began to drag it toward the baobab. The dead tree was heavy, though, and its branches kept digging into the ground, making it difficult for William to move. I was certain that William intended to make a ladder that would enable him to reach the first branch of the baobab. He struggled and pulled but made little progress. Finally, he seemed to give up and sat looking at me. I felt his idea was worth a little assistance. I did not want him to get discouraged. He panted eagerly as I got up, then ran to the base of the baobab and waited expectantly.

Being taller than William, I was able to lift as well as drag, which made it much easier to get the skeleton tree to the baobab. William helped me prop it against the trunk. It did not quite reach the first branch, but William climbed as high as he could, then jumped skillfully, catching the smaller branches that grew from the first main branch. For a good half an hour he remained in the tree and fed well on the fruit.

It was not long before we found another baobab tree even more impossible to climb. It was laden with last year's fruit, large, dark-brown, velvety balls unattainable to anything that might feed off them. This particular baobab was tall as well as wide, and its first branch grew too high for William to find a natural ladder that would reach it. There was a kenno tree growing beside it, but just too far away for the branches to interlock and provide a passage. It seemed that the huge feast of fruit was going to hang there till it fell with the wind or went rotten. William asked me several times to help, but there was nothing I could do.

As a last measure Julian and I flung rocks at the fruit. Half an hour later my arm and side ached but I did not have a single fruit to show for my efforts. Julian had better luck—he hit and brought down one fruit, which William and Pooh shared. When William had finished his half he sat and gazed up at the rest. Then he reached out and picked up a rock the size of a cricket ball. I knew his underarm throw would never come near the fruit he wanted. To my amazement William seemed equally aware of his limitations. He did not even try to throw from ground level, but climbed into the neighboring kenno tree until he was at the same height as the fruit. At the end of a small branch, only a few away, hung two fine specimens. William placed himself directly opposite them, swung his arm back and forth three times, then hurled the rock. Sadly it missed the fruit, but hit the baobab trunk with such force that the rock shattered. William seemed to realize that his enterprise was impossible and did not attempt it again. Instead he continued to hand me rocks. I felt sorry for him; he had tried so hard that he deserved a reward. Then I had an idea. I wrestled with my conscience for a while, but finally decided that I would put it into practice.

The following morning Julian, William, Pooh and I set out for the baobab tree. Julian carried a coil of rope over his shoulder. When we reached the baobab, Julian climbed into the neighboring kenno tree, still with the rope on his shoulder and with a rock in his pocket. Our plan was to attach one end of the rope to the kenno tree, slinging the other, weighted, end over a baobab branch and letting it hang to the ground, thus providing a single-strand rope ladder that I thought the chimps could climb. After several attempts Julian managed to throw the rope over the bough of the baobab. It took William thirty seconds to realize what Julian had done. He ran to the rope and pulled it, then climbed about three feet off the ground. The damp rope was slippery, and it stretched. William did not trust it.

He climbed down. Holding the rope in one hand, he walked round to the other side of the baobab tree and tried walking up the trunk with his feet while climbing hand over hand on the rope. The rope was in the wrong position for that idea to work. William slid the few feet down the baobab tree and sat on the ground, the rope still clutched in his right hand. He sat calmly for just under a minute, then got up, still holding the rope, and walked to the base of the kenno tree. Almost without hesitation he climbed toward Julian and chucked the rope underarm in his direction. It caught on a branch, and Julian leaned forward and picked it up. William had

William on the rope walk

just shown us that if we tied both ends of the rope to the kenno we would have a bridge that would be twice as strong as the ladder, many times shorter, and much easier to climb across. I would have blushed at my own stupidity if I had not been overwhelmed by William's behavior. Julian dutifully pulled the rope taut round the baobab bough and attached it again just above the first knot.

William immediately approached the double-rope bridge. He fingered the end for a few seconds, then decided to try the rope. It stretched beneath his weight and he leaped back. He sat for a few more seconds, then tried again. He kept his feet on the bottom rope and held the top rope with one hand and a kenno branch with the other. Slowly he edged out. He hung onto the kenno branch for as long as he could, then made a lunge and grabbed a small, leafy baobab branch. In a flash he reached the main bough and climbed into the tree. He food-grunted excitedly and immediately picked a fruit. It was extremely satisfying that among us all we had gained access to the wealth of the baobab, even though I knew we had resorted to methods that the chimps would never be able to use when they were finally alone.

When William smashed a baobab fruit against a bough or a rock, it was with all his strength and concentration. He was bent on getting at the food. Failure to achieve something he wanted quickly frustrated him. Pooh was more easygoing; his motto was if at first you don't succeed, play, try again, rest, play, try again. He usually succeeded in the end and, unless he was desperately hungry, he did not mind taking his time. Tina, when she was with us, worked like a machine. She was a skilled baobab opener and usually held the fruit at the end of its long stalk. With effortless-looking swings she would crack the fruit the first or second time it hit the bough. Then she would insert the tip of her long canine teeth into the crack and pry the fruit open with her hands.

One afternoon, when Pooh and William had eaten their fill of baobab, William climbed to the ground, holding the stalk of a fruit in his teeth. He lay near me for a few minutes and groomed my leg, leaving the fruit balanced on his belly. When he got up he walked slowly to a thicket, the fruit still dangling from his teeth. He seemed to doze, then must have wanted more to eat, for he took hold of the fruit and whacked it hard on a branch. He examined it and found a hairline crack along one side. He tried to insert his teeth in the crack without success. Growing above William's head was the branch of a caramel bush, and close to each leaf was a long, sturdy thorn. William reached up, pulled the caramel branch toward him and with his teeth detached a thorn. Carefully he took the thorn out of his mouth, lifted the baobab, found the crack and tried to insert the thorn into it. The thorn buckled and snapped as William pushed. He picked another and tried again.

The baobab fell to the ground. From the sound it made on the rock, I suspect the crack widened slightly. William examined the crack and inserted his lower front teeth in it, then pulled downward with his hands. It was a tough shell: his teeth slipped and his lower lip was pinched in the crack. I winced involuntarily, imagining how painful it must be, but William scarcely flinched. He picked a third thorn and wedged it in the crack; playing with the thorn, he pulled it out, then wedged it in again. Finally, using his teeth in the same way as before, he opened the fruit and lay back like an emperor, the split fruit on his stomach, daintily nibbling small pieces, sucking and savoring them lazily. By this time I was sitting on a branch next to him trying to photograph the proceedings. He paused and idly watched my movements, then took a piece of powdery, white baobab out of the shell and graciously held out his offering. I accepted gratefully. I was surprised and touched at his generosity.

22. FACE TO FACE

For over a week Tina had been with us each day and had nested close to camp in the evening. One morning I woke to hear chimps pant-hooting. The calls seemed to be coming from the stream where I washed each evening, only about two hundred yards from camp. I quickly realized that I did not recognize the voices; it was not Tina and William, it must be wild chimps.

Pooh was still fast asleep on the platform, and William and Tina, I presumed, in the gully. It was scarcely light. I set off immediately with Pooh. The chimps were not at the "bathtub," as I called that section of the stream—they were farther down, but the calls indicated that they were moving toward me. Pooh and I hid ourselves, Pooh still so sleepy that he sat on my knee and dozed.

Julian found me about a quarter of an hour later. William and Tina were with him. The calls had ceased and I wondered if perhaps the chimps had left the stream and gone elsewhere. We waited another few minutes. Pooh followed Tina and William down the slope. I could not see them all of the time, but I could hear Pooh playing. Suddenly Tina pant-hooted. She was down by the stream about twenty yards ahead of us, but because of the grass and vegetation, I could not see her. A chorus of pant-hoots echoed Tina's. She began to make submissive cough sounds, then broke into a scream. She was running toward the plateau, parallel to me. I scuttled through the grass and reached the plateau in time to see her cut down into the gully behind the camp.

Three adult male chimps headed out onto the plateau just ahead of me and pursued Tina. The first two were in prime condition—one had a brown face and an abundant glossy coat. The third chimp was slower than the other two and seemed a lot older. When he had passed me, I followed quickly. Pooh leaped onto my back. I didn't know where William was. I ran quietly along the edge of the gully.

Tina was still screaming. When I was directly opposite camp on the other side of the gully, the first two chimps attacked Tina. Her screams changed from long drawn-out cries to the shorter, differently pitched screeches that mean an attack is taking place. Tina managed to break away and ran up the slope toward camp. The two wild chimps followed. By now I was desperately worried for Tina, so I stood up on a rock in full view and pant-hooted. The wild chimps stopped in their tracks and spun round. Tina continued screaming as she dashed into camp. The wild chimps seemed shocked into immobility by the sight of me standing on a rock with Pooh on my shoulders. Pooh followed my pant-hoot with two aggressive whaas.

Suddenly William whaaed too. He was in the gully, hurrying up the stream toward the other chimps, closer to them than I was. The spell broken; the wild chimps raced up the slope. They ran into camp, then swung round again and disappeared along the camp side of the gully. Tina was in the vines behind the shack. She panted frantically when I approached, then quickly licked the blood that was dripping from her elbow. She had a long gash above her right elbow and a smaller wound on the outside of her right hand. She seemed dazed but concerned about the blood, and fastidiously licked herself, sucking away any signs of it from her coat. When she had cleaned herself and her wounds had ceased to bleed, she lay down on a bough to rest.

To try to console her I offered her some onions, which she usually loved, but this time she refused. Instead she left her resting place and went down into the gully. Her injured arm was kept close to her body and she walked slowly but with a surprisingly upright stance. I hoped she had not been hurt more seriously than had appeared at first. William walked slowly beside her. When they reached a small clump of trees halfway down the gully, William climbed into one of them. Tina followed, using only three limbs—she hugged her injured arm to her chest. Once in the tree, Tina licked and inspected her wounds again, then began to eat leaves. She fed for a few minutes, then rested again, occasionally cleaning her hand and her elbow. Suddenly she sat bolt upright and stared into the gully. Then she turned to William and held out her hand to his mouth in a request for reassurance. Her lips were drawn back nervously, exposing her teeth. Tina was afraid.

William moved toward her and followed her stare. Slowly his coat rose till he was bristling. He swung quickly from the tree and disappeared downhill. All at once I heard a short scream. I couldn't see what was happening. It wasn't William's voice. I could still see Tina, and Pooh was hurrying down the slope just ahead of me. The scream was followed by hoarse, frightened panting—the sound a nervous chimp makes when it submissively greets another.

Unexpectedly there came a loud, surprised whaa bark and the noise of several chimps screaming. Tina hurried into the gully. The screams were of fright, not of attack. I suspected that the wild chimps had seen Réné and Julian, for after the screams there was silence. I strained to watch them climb up the other side of the gully, but saw nothing. About ten minutes later Pooh came scampering up the slope to sit near me. I could not make out Tina or William, and the gully was silent. I crept down carefully and scanned the opposite slope with my binoculars. At the top of the gully I could perceive the silhouette of a chimp gazing down at us. It was only William. In some lianas near him I saw Tina. I walked toward them. I would dearly have liked to know what had happened in the gully. I wondered how many strange chimps had been present—two or three, I guessed—perhaps females. They were more likely to have made the panting sounds at the sight of William.

At Gombe I had heard of cases of hostility between members of two neighboring communities. I had also learned that there were some individuals who seemed to have a kind of immunity and who traveled back and forth between the two groups undisturbed. Young females from one community sometimes transferred into the other and remained there. Niokolo Koba has a much lower chimpanzee population than Gombe and, according to the limited information I had at the time, it seemed the small Mount Asserik community had a much greater feeding range than the communities at Gombe. I hoped that because of this, the Niokolo chimps would be less possessive of their territories and would tolerate strangers more easily. William and Pooh were still too young to pose a threat to the male hierarchy, and as Tina was a young female, I expected her to have little problem in being accepted.

During the rainy season, when fruit was abundant, the wild chimps came together in bigger groups. In such groups they would almost certainly be more excitable. Perhaps that was why Tina had been attacked by the three males that morning. For all I knew, the three attackers might just have been bullies in a bad mood. The wild chimps in the gully, on the other hand, did not seem to have been aggressive; in fact, judging from their screams of fright, they had been positively submissive.

TINA'S WALKABOUTS, when she disappeared for days, became less frequent. She remained with Pooh and William and followed us each day on our long walks. There were periods, even in the rains, when there seemed to be no fruit around. Some species had finished fruiting, and there was an interval before others became ripe. Tina quickly taught us that there was always an abundance of food—many different leaves and grasses, flowers and bark, some of which even I found very palatable. She introduced us to seven different varieties of edible seeds and taught William and Pooh that baobab fruit could be eaten even when green.

Tina also taught William and Pooh that vegetable matter was not the only source of food. As she passed a sprawling bush she paused briefly, but not to feed; instead she picked a long, thin, green stalk and held it in her mouth, then stripped the leaves from the stalk by pulling it through her closed hand. One or two leaves remained at the tip. She nipped these off with her teeth.

I held my breath as I watched Tina walk purposefully to the base of a termite mound. With the nail of her forefinger she flicked away a small pile of damp-looking soil and exposed the opening to the termite passage. She bit a tiny piece off the end of her stalk and pushed it efficiently into the hole. Almost immediately she pulled it out—there was nothing attached to the stick. Ten times she pushed the stalk into the hole without achieving anything. She moved round the mound and opened another passage, bit another piece off her stalk and pushed it into the termite mound. The sixth time she withdrew the stalk there were two large termites clinging

Bark eating

to the end. Tina quickly lipped them off and chewed them. From then on there seemed an endless supply of termites to be fished from the passage. She worked quickly and efficiently, drawing her fishing tool across her wrist each time she extracted it. Some of the termites came off on her wrist, but they had to crawl through her coat before they could bite her with their powerful pincers, and this gave Tina plenty of time to eat those on the stalk, then deal with those on her wrist. What struck me most forcibly was that Tina's method of extracting termites was exactly the same as that used by the Gombe chimps thousands of miles away.

Whenever Tina bent her stalk while inserting it into the mound, she nipped the bent piece off. Eventually her tool became too short to use effectively, so she strode off, selected another suitable stem, stripped the leaves off and came back. Pooh picked up the abandoned stalk and tried poking it into any small openings he could find. Eventually he gave up. I wanted Pooh to learn to fish for termites, and decided to try to teach him. I walked to the same bush Tina was using, broke off a stem and stripped the leaves from it exactly as she had. Pooh watched me with growing interest. Then I tried Tina's first hole, imitating the way she held the stalk. After several attempts I pulled out the twig to find two termites clinging grimly to the end.

I picked one off, pinched it to render it harmless and offered it to Pooh. He looked very skeptical, despite the "good-food" grunts I made to try to persuade him. There was only one course of action left—I had to pop the wretched termite in my mouth and chew it, all the time making the appropriate grunts of appreciation. I expected it to taste foul, but surprisingly it didn't taste of anything much. I pinched the second termite, popped that one in too and continued fishing like Tina. After a while I got Pooh to taste one, but he did not appear enamored with the flavor and spat it out. There was a limit to how many termites I was willing to eat in order to persuade my reluctant pupil, and finally I gave up, disappointed with the results of the first lesson.

It seemed I had achieved very little. Tina continued to hunt termites for thirty-five minutes, and just before she stopped, I noticed Pooh had taken my stem. As the hole had only recently been worked, he got some termites almost immediately. He sat there staring at them, not sure what to do next. Then he tried to detach one with his fingers. The termite released the stem and gripped Pooh's finger instead. Pooh leaped up with a surprised scream and tried to shake it off his finger, but the insect clung on. Pooh quickly scraped it off on the ground and, not surprisingly, lost interest in termite fishing for the moment.

Réné and Julian showed us three types of roots that were good to eat. William and Pooh soon learned to recognize these plants, and I taught them to dig up the roots. But one had to dig fairly deep and carefully, and William usually lost patience halfway through and would pull furiously on the top of the plant. The stem would then tear away, leaving the root still safely embedded in the soil. At this point the chimps would more often than not drag me to the plant, give me a stick to dig with and ask me to get the root for them. I would oblige with enough roots to whet their appetites, then tell them to get on with the job alone. Occasionally they succeeded; usually they didn't. The roots were delicious, but Pooh and William seemed to think it was an awful lot of hard work to get at a relatively limited food supply. However, I felt that they had learned about one more food source they could fall back on in times of

Fishing for termites

need—food that none of the other animals seemed to know how to obtain.

William and Pooh also learned to eat mushrooms from watching us collect them. I was worried they might think all mushrooms were good to eat, but they ate only those they saw us pick. Tina would regard them intently. She got as far as picking a mushroom, smelling it several times and then tasting it, but she didn't seem to appreciate the delicate flavor. Tina was very hesitant about accepting anything new, foods especially; she was never interested in the roots, for example, and it was only after several months of watching Pooh and William eat rice that she finally began to eat it too. I was happy that I had released Tina when I had, while she still remembered her life in the bush and was young enough to make changes easily. A young chimp will readily imitate someone he respects and looks up to. Once he reaches adolescence he is far less impressionable. Fortunately Tina was older than Pooh and William and commanded their respect; this made her a very effective teacher for them.

THE GREEN, DAMP DAYS of the rains grew into weeks and then months. The chimps were healthier and more content than they had ever been, and they were slowly acquiring a sound knowledge of their new home. They always had the camp as a secure base from which to explore and this I still felt was essential. They learned so much more easily when they were confident and relaxed, and as there was no desperate pressure on them to survive, William and Pooh absorbed their new way of life without being aware of it.

Without their really noticing, their cups of tea became fewer and fewer and they learned to find long drinks of cool water from the stream just as refreshing after a day out in the bush. They got tougher and more alert and came to rely on the valley for food. Occasionally, if I thought Pooh had not eaten enough during the day, I would give him a meal when I ate in the evening, but William was entirely self-sufficient and seemed to be growing at an alarming rate. He still did all he could to steal from camp though—not because he was particularly hungry, but because camp food was a luxury which he enjoyed.

During the later part of the rains we had a huge crop of grape-like fruit in our valley. They were oval in shape and hung in clusters from the trees. When ripe they were golden yellow or orange. They were juicy with a strong, slightly sour flavor, and the chimps loved them. During the weeks when the trees fruited, the valley was a Garden of Eden for the chimps. Near the waterfall was a grove of about ten of these trees, all laden with fruit. It was an intoxicating sight, and each time we approached, Tina, Pooh and William would pant and hug each other in excitement, and hurry toward the trees, food-grunting frantically.

One day Julian and I sat in the grove while the chimps feasted in the trees. They had been there for an hour when Tina climbed down and vanished silently into the valley. Her behavior puzzled me, so I got up. Then I heard wild chimps pant-hooting, but they seemed a long way off, probably at the base of Mount Asserik. Pooh and William were still in the trees, but looking out intently over the plateau behind us. Both of them had their hair erect, and I wondered if perhaps they could see chimps on the other side of the plateau. Suddenly there was a tremendous chorus of pant-hooting right above me on the plateau, and the soft thud of running feet. William and Pooh quickly but quietly climbed to the ground, and William disappeared in the same direction as Tina. Pooh stood a few yards ahead of me, staring up to where excited food-grunts were coming from. I heard the grass rustling and knew one of the wild chimps was approaching the grove.

There was just enough time for Julian and me to hide behind a tree before a young female chimp, just a little smaller than Pooh, entered a clearing below the fruiting trees. She had a pale, flat little face with slightly slanted oriental eyes that reminded me of Wong. She stopped a few yards in front of Pooh, looked at him, then walked straight toward him. Pooh didn't move. At that instant an adult female came into the clearing with a young infant clinging to her belly. She was closer to me than either Pooh or the small stranger, so I had to duck back behind my tree and miss what took place between Pooh and the young female. The mother chimp walked past me, then turned and grinned nervously in Pooh's direction. I tried to keep absolutely still, but she turned her head round a fraction farther and we looked into each other's eyes. She bolted, running back the way she had come. Pooh gave a short, startled scream as she tore past him. The young female followed her. I almost cried. I had spoiled Pooh's first chance to meet wild chimps. They had certainly seemed friendly toward him, especially the young female.

We went up onto the plateau with Pooh and William, hoping that if the wild chimps were watching, we would present a harmless picture. Then we returned to the grove. I was almost certain that I had scared the chimps away, but just in case, Julian and I sat hidden in a tangle of bushy vegetation. A path ran through it onto the plateau, and if the chimps appeared again and accepted Pooh and William, Julian and I could move farther away unseen.

At first Pooh and William stayed in the blind with us. Some more wild chimp calls told us that the others had not gone far. William moved out and began to eat grass stalks. Pooh sat on a dead log, idly picking off the bark. An hour later Pooh climbed into one of the fruit trees and began to feed, but William, obviously full up, remained on the ground. Suddenly just behind me I heard chimps food-grunting. William came to me, his coat bristling. His lips drew back nervously and he held my hand, looking from me to the first chimp that came into sight. Just below the blind we sat in there was a grape tree full of fruit. A line of chimps, about five of them, were walking along the fallen tree through the tangle of vines. All were food-grunting madly, and some of them began to scream with excitement. Pooh began to scream too, caught up in the atmosphere of it all. The first female chimp saw him, and the screams of excitement got louder. William moved out into the open, and to my joy responded beautifully. He was grinning submissively and flexing his elbows in a tractable, bobbing way.

At least one chimp was approaching William, who continued to behave respectfully and submissively but stood his ground. Suddenly one of the chimps in the vines, a youngster, began to scream as if he had been threatened by another chimp. Pooh, carried away with the general excitement, gave a hostile whaa bark. All the wild chimps, now about ten of them, began to whaa back. William and Pooh insolently whaaed in turn, confident that I was close by to protect them. I began to tremble—the noise was terrific and some of those chimps looked enormous. I could hear the screams get more and more frenzied and was certain there was going to be some sort of action.

The blind began to feel like a cage. We were too close to such a large excited group of chimps; if we were attacked we'd be easy to catch. I told Julian to move out slowly, keeping the bush between him and the wild chimps, and to sit in the long grass a few yards away. The noise around us was now deafening. I waited for Julian to get clear, then began to climb out too. Before I could turn round to pick up my bag and plastic raincoat, I heard the first charge coming down the path that led into the blind. I scrambled out and got to my feet just in time, for what seemed to me the biggest chimp I'd ever seen sped into the blind and out the other side. He stopped abruptly less than three yards in front of me, and rushed back the way he'd come. I was terrified—my knees were literally knocking. Pooh and William were sitting beside me holding my trouser leg. William climbed onto a low branch of a tree just behind me. Pooh still clung to my leg. I put my hand down and pulled him into my arms. Almost immediately there was wild pant-hooting and thudding of feet as another chimp careered down the path through the blind, stopped just in front of me, then ran back again.

William was still whaaing loudly behind me. The other chimps were still screaming, but I felt sure that only the two which had run down the path and come face to face with me knew I was there. The chimps all around us seemed in a frenzy of excitement. A few seconds later I could hear more branches being torn off, and a third bristling adult chimp displayed down the path, dragging a branch. Directly behind him came another, but as they cleared the blind, the sight of me standing a jump away with Pooh in my arms halted them in their tracks. They both ran back along the path. Suddenly the cries abated, and the sounds of fleeing chimps replaced them. I looked up to see three of them standing on the edge of the plateau peering down at me, but at that moment they disappeared.

My knees felt so weak that it was difficult to walk, but I wanted those wild chimps to get a good look at us all and to see that we meant them no harm. I scrambled up the slope with Pooh, while William followed closely. When I got to the plateau I sat on a rock in the open and put Pooh down beside me. William climbed into a tree nearer the edge of the gully. Soon after I sat down I saw one chimp peering at us, but he quickly left. If only Pooh hadn't whaaed like that, things might have turned out differently, but the second chimp encounter of the day had been far from amicable. I felt we had a small war on our hands now.

Next day the chimps and I returned to the same area, accompanied this time by Réné. It was almost four o'clock when I heard a young chimp scream just behind the grove. Several chimps appeared in the stream bed, and some climbed the tree that William had fed in earlier. Others walked up the slope to the grove. It was difficult to know how many were present, for the grass hid them for much of the time. William sat just behind me, Pooh to my left. Pooh was excited, his coat was partially raised and he was going through the motions of clapping his hands without actually making any sound. Three chimps—an adult male, a female with a slight swelling and a juvenile—were feeding in the tree. Pooh relaxed and watched. There seemed to be other chimps in the grass below, but I couldn't see how many or what they were doing.

One of the chimps began to walk out along a branch, obviously bent on the huge and inviting bunch of orange grapes that hung temptingly from its tip. With a dreadful cracking sound the branch broke and fell to the ground, taking the chimp with it. Pooh clapped his hands. I held my breath and stared anxiously round the boulder to see if the chimp that had fallen was hurt. The others continued feeding, unconcerned for their unfortunate friend. The grass was too high for me to see the chimp on the ground—there was no sound, so I assumed he was either unhurt or knocked unconscious. It had been a pretty long fall, twenty feet or more. Suddenly there seemed to be a squabble in the grass at the base of a tree, and I caught a glimpse of a female with a large pink swelling running up the slope to the grove screaming her head off. Pooh whaaed. I cringed back behind my rock, expecting that they would notice us at last, but the screams must have masked Pooh's voice, for the next time I peeped out the chimps in the tree were still feeding.

To my astonishment William suddenly passed me and made his way down the slope toward the wild chimps. As he approached he began to bob up and down and make the correct submissive sounds. He was about fifteen to twenty yards from them when, for some reason that I couldn't see, there was an outburst of excited screaming and all the wild chimps hurried away up the slope. I was certain that neither Réné nor I had been seen. Was it possible that the wild chimps associated William and Pooh with humans and were afraid of them? William stood with his back to me and watched the chimps flee, then sat down heavily and stared up the slope at the grove.

I waited half an hour, then crept down to the tree in case the chimp that had fallen was still lying in the grass. The branch was broken in two places but there was no sign of the chimp. Wild chimps were calling at intervals, but were already some distance away. I was disappointed that the chimps had run off so suddenly, but also excited that William and Pooh had had the chance to watch wild chimps at such close range. I was particularly pleased with William for voluntarily approaching the small group. I wondered how and where he had learned such good manners. He had so rarely needed to be submissive during his life and yet had reacted perfectly to the wild chimps.

An hour later, ready to go home, I began to put away my notebook. By chance I glanced at the grape tree. A large adult chimp was silently climbing the trunk. Five adult chimps, at least three of them strong, husky males, were feeding in the tree. Pooh and William did not notice the newcomers till one of them food-grunted, then both watched intently. Pooh and William watched silently for a while, then Pooh came to me and sat with one arm round my waist.

The wild chimps fed for about a quarter of an hour and then one by one descended into the grass, where I lost sight of them. Five minutes passed without a sound, and I thought they must have gone. I inched out from behind my rock to have a look. Two chimps were walking slowly up the slope toward us—a juvenile about Pooh's size and an adult. Heart pounding, I crept back behind my rock. William began to bob and grin again, squeaking and coughing his respect and submission to the strangers. He moved past me toward the wild chimps, then stopped about six yards in front of my rock. I soundlessly willed him to move farther down the slope; otherwise we would be discovered.

Pooh walked out and sat by William—silent, with his coat partially fluffed out, staring down the slope. William's bobbing became more frantic. He desperately wanted to move forward but seemed too nervous to do so. He kept quickly glancing back at me and then ahead of him. I could not see down the slope, but from William's eyes and his behavior guessed the chimps were approaching. I was shaking with excitement and prayed that this time William and Pooh would be accepted. Then I heard submissive coughs and excited squeaks from a wild chimp, on the other side of my rock. The suspense was terrible. William and Pooh had made contact with a wild chimp. There was no screaming or aggression, just the small, energetic noises of strange chimps meeting. I peeped out. Instead of the juvenile and what I presumed to be its mother, five adult male chimps were on the slope, each with his coat partially raised. One was standing inches away from William, a second was walking past them both to where Pooh sat. The other three were coming up the slope at a walk, looking nervous. I didn't dare peep for too long in case I was seen. William was still being submissive and doing a lot of high-pitched squeaking. The atmosphere was tense, I knew, but surely if those chimps had wanted to attack, they would have done so by then.

Pooh got up and moved beyond Réné's rock. He was slightly ill at ease, though not sufficiently afraid to seek reassurance from his human protectors. I hoped the wild chimps would lead off across the slope, but the seconds ticked by and they remained just past the boulders. Then, directly in my line of vision, about six feet away from where I crouched, a muscular, exceptionally good-looking male chimp climbed slowly onto Réné's rock. His coat was not as sleek as it might have been normally, but he certainly didn't look aggressive—merely uneasy.

I held my breath and waited for him to spot me. The chimp moved higher on the rock, and was about to leap to a neighboring one when Réné looked up. The sight of so powerful a wild creature just above him would have tried even the stoutest of hearts. Réné and Julian had both felt the fury and strength of William, and William was half the size of any of these chimps. Réné started, then backed off quickly. The wild chimp jumped violently, stared at us both for a fraction of a second, then ran down the slope. His companions did not hesitate to follow him. William, to my surprise, went after them at a fast walk, looking round nervously for the cause of their apparent fear. Réné and I had moved out of hiding and now sat in full view on the top of a flat rock. Pooh was clinging to my waist. Like Pooh the previous afternoon, William could not comprehend the sudden flight of the wild chimps. How could I expect William to understand that people whom he had grown up to trust and depend upon could prove such a terrifying spectacle for the wild chimps?

Reviewing the evening's events in my mind, I felt more and more elated. William and Pooh had been in the midst of wild chimps. Réné told me that he had seen one of the males touch William's face. He said that though their coats had been partially raised when they were a good distance away from us, they had all calmed down when they reached us.

Next morning we made our way back to the same place. The first things to catch my attention were four fresh nests in the tabbo tree right above the spot where the wild chimps had discovered Réné and me the evening before. They must have been built after we'd left, during the last hour of daylight. A little more searching and I found a fifth nest in another tabbo, just down the slope. Perhaps the chimps had not gone far after all. Perhaps they had watched us leave. Perhaps we had not frightened them as much as I had thought. I could not be sure, for it could as easily have been five different chimps that had chosen to nest there that evening.

Beneath each nest I found a fresh sample of dung. It was possible the chimps had seen us arriving and had left only minutes before. With a twig I sifted through the dung to see what they had been eating. It consisted almost entirely of the grape-fruit stones, but in one sample there were small pieces of what looked like baobab leaves and long strands of fibrous baobab bark. During our walks we had often seen baobab branches almost totally stripped of fresh bark. It was good to know that Tina was really teaching William and Pooh the diet of wild chimps in Niokolo.

Julian and I decided to spend the morning beneath the clump of small trees covered with vines on the edge of the plateau—at least there we could get out of the way without being seen if the wild chimps came up the slope. By then we had been sitting in the blind for an hour. William and Pooh were close by, eating the pith of some grass stems, resting and playing. There was a sudden short distress cry from a young chimp halfway down the slope, but to the waterfall side of the blind, near the grove of trees with grapelike fruit. Then I heard a branch being broken.

William climbed onto the plateau, stepped round the blind and walked down the slope in the direction of the feeding chimps, making the pant-cough greeting. I strained hard to catch each sound and to picture what the situation must be. There was the excited, high, screaming sound of a nervous chimp, and I thought that William had been seen or had even approached the strange chimps. The squeals sounded like a female, or a young, unconfident chimp, reacting uneasily to the presence of a stranger. Pooh listened with interest but remained near me.

Suddenly there was a burst of excited, aggressive screaming, and William ran onto the plateau. He sat down and grinned nervously in the direction of the blind. No one followed him up onto the plateau. William sat for only a minute or two, obviously ill at ease, then walked out of sight toward the chimps again. There was silence for ten minutes, and I thought the chimps had fled. Then William began his respectful greeting coughs and pants again, and another bout of screaming exploded. This time I heard a chimp slap the ground several times and imagined one of them had displayed. Again William ran out onto the plateau and approached the blind, looking at me through the leaves and grinning. I threw him a kiss and told him not to worry. They weren't chasing him, they were just rather apprehensive. I was sure it was my fault—they associated William with me.

William entered the blind and rested. Pooh played noisily, rolling a large rock around in a ground nest he had built. It was two hours before William ventured out onto the plateau again. During this time I had not heard any chimp vocalization, but the occasional snapping of small branches indicated that the wild chimps were still present. I was amazed that they had remained, with all the noise Pooh had been making in his nest. William was hesitant at first. He walked along the edge of the plateau, past the place where the chimps were feeding below, and climbed into a fig tree about twenty yards away from the blind. I pointed out to Pooh that William was feeding on ripe figs and urged him to do the same. He obligingly sauntered over to the fig tree and climbed it. William was now glancing frequently down the slope. As Pooh began to feed, William started his pant-hooting. It seemed a chimp was approaching.

Pooh gave a nervous grin and a squeal and hurried out of the tree. William followed. Pooh came right back inside the blind; William stopped about ten yards from it, then went back again, bobbing and making the usual sounds—obviously very nervous. He reached the edge of the plateau and retreated slowly a few yards, bobbing all the time, squeaking, grinning, torn between a desire to meet this invisible stranger and doubts about his reception. Three times William approached and retreated; then, as he went forward the fourth time, another chimp began an excited, nervous squealing. Still I could not see it. William kept looking at the blind—I parted the leaves so that he could see my face and know I was still there. Six times in all William went forward and retreated. On the sixth occasion he kept on backing away step by step, and from his behavior I knew the wild chimp was walking toward him. Fortunately William was backing out onto the plateau and not toward me.

An adult chimp, his coat partially erect, was closing in on William. I really couldn't tell how the chimp was going to react. He didn't look friendly, but neither did he look particularly aggressive; rather he seemed wary and unsure. William stopped, still bobbing though. Then the wild chimp did a strange thing—he seemed to put his face right up against William's. William grinned and squeaked but stood his ground. A second male chimp walked out onto the plateau and stood glaring at William and Pooh—or was it past William and Pooh toward the blind? The first male then did something even more curious. He moved cautiously past William, made a slight detour, then stopped and carefully approached the edge of the plateau behind the blind. Pooh leaped down and entered the blind. I froze. Not again, please God, don't let me spoil it for them again. My bag, sparsely covered with leaves, was lying at the entrance to the blind. I thought the chimp was going to sniff it, but he merely stared at the spot, then bent his elbows and, crouching, examined the blind. His coat was still partially erect.

The opening of the blind was partly covered with vines, so I wasn't sure if he could see us. He stood up abruptly and walked back past William, who was much calmer, and into the fig tree. The second wild chimp, looking even more anxious than the first, followed him. William walked into the blind. The first chimp picked at some figs, but he was far from relaxed and swayed several times while staring around him. The two chimps remained in the fig tree three minutes, then descended and walked out of sight the way they had come. I hugged William several times and whispered my praise of his courage and persistence. He looked pleased and patted my hand affectionately.

Shortly afterward William and Pooh came back to the fig tree and fed. At first William stared down into the valley, and for at least twenty minutes I could hear the sounds of the other chimps feeding about fifteen yards away. Then all was quiet. William and Pooh remained in the fig tree half an hour, then returned. On the way back William stood upright and looked down into the valley briefly. During the remaining few hours at the blind a group of chimps called at regular intervals from the waterfall, then seemed to move farther away. They did not come back to the fruit near the hide before we left to go home.

23. BOBO

IT WAS OCTOBER—six months since we had left Abuko and made Niokolo our home. The chimps were slowly and naturally growing away from me. There had been no sudden partings or hardships for them. The process had been so gradual that I don't think they even noticed it. Again I thanked God for Tina. How much more difficult it would all have been without her! More and more often now I would go away from camp without the chimps, leaving them to lead their own lives in their own way.

Toward the end of September I had been expecting a visit from Nigel and Daddy, but weekends passed quietly with no sign of anyone. I decided to risk going to Niokolo to see if there was any mail or news of an impending visit. It was a nerve-wracking twenty miles; the grass obliterated the road, which we left on several occasions—fortunately never hitting boulders large enough to damage the Land Rover. All the slopes had been deeply eroded by the rain and there were several long boggy patches that Felicity only just negotiated.

A stack of mail—the accumulation of months—waited for me, including a telegram a couple of weeks old from Italy. An Italian girl named Raphaella Savinelli had cabled to ask whether she could bring her three-year-old chimp Bobo to my camp. He had lived with her for two years, but it had become impossible to keep him any longer. There were also letters from Dr. Brambell in London. The arrangements for Yula and Cameron were finalized. British Caledonian Airways had agreed to give me a ticket to go to England in January to pick up the chimps, and we were all to travel back together on the British Caledonian flight direct to Dakar in Senegal.

The rains eased and finally stopped. Mount Asserik changed

from flourishing green to its mellow autumn colors. The grass of the plateau looked like a huge field of ripe golden corn, and many of the trees changed to yellow, then orange, and finally shed their leaves. It was a completely different landscape from that of the rainy season, but equally beautiful. Soon the park would open to the public again.

One day I had driven out for a few hours to carry some supplies to a gang of laborers who were working on the park roads a few miles away. As I drove back I noticed a huge cloud of black smoke billowing up into the sky near Mount Asserik. It was the first of the bush fires. The closer I got to camp, the more anxious I became about the smoke—it seemed precariously near camp. Finally, when I reached the junction, I found the ground charred and smoking. The fire was roaring its way up the steep slope to the plateau and to camp.

It would be the first big fire the chimps had experienced, and I wanted to be sure they had the sense to keep in the gully or down by the stream—places that were unlikely to burn. They might panic, I thought, and get trapped in the fire. I had to get to camp fast. I drove the Land Rover as far up the slope as I dared. A line of flames barred the road. On the front of the car were two spare jerry cans of gasoline held in place with special brackets. I put the two jerry cans in the back, then broke off as big a green branch as I could easily wield. The dead grass, dried by the sun, provided the fire with ideal fuel, and it roared and sparked quickly up the hill ahead of me.

The heat and smoke were almost unbearable, but I finally managed to extinguish the flames across the road. It would only be minutes, though, before a flying spark ignited the road again. I ran back to the car. It was difficult to get up speed on the steep hill; the wheels spun, throwing up black gravel and ashes, but then gripped, and the Land Rover screamed through the small passage I had made in the wall of flames and up ahead of the advancing fire to the plateau.

Réné and Julian had already begun to cut a firebreak around camp. Pooh was with them, but Tina and William had been out of camp since I'd left that morning. There wasn't time to cut an effective firebreak; the flames were already visible in a long, unbroken line stretching right across the far end of the plateau. Julian and Réné cut a stack of green branches and brought them to camp. Then

Bush fire on the plateau

for half an hour we burned the grass around the camp, beating it out as soon as we had made a border wide enough to protect us.

Once the camp was reasonably safe, I had time to worry about William. Tina, I knew, would have already lived through a bush fire; I hoped William would keep close to her and heed all she did. I held Pooh who, once secure in my arms, peacefully watched the spectacle. The firebreak worked well. The fire roared, cannoned and spat all round us, then passed on, leaving us hot and smoky but undamaged.

The plateau, which only an hour before had been a swaying sea of golden-orange grass, was now a charred wilderness, as bleak as any scene from the moon. The vegetation and light forest around the plateau looked brown and drooped miserably. It was a depressing scene. The dry season was here. Until the next rainstorm, seven months away, most of Niokolo would take on a barren appearance, another, much starker kind of beauty. Only the hidden valleys would remain green to hint at the lushness of a rainy season. The fire was still clearly audible, crashing its way up the slopes of Mount Asserik, when Tina and William came out of the gully into camp. Neither seemed particularly perturbed. William sneezed frequently, but that was because of the smoke that still hung low in wafting clouds over the gully, trapped by the interlocking branches of the forest.

We passed the rest of the day in camp. William and Tina spent most of their time feeding on kenno leaves and the black berries of the kutofingo bushes. Pooh seemed relieved to have me around and became quite skittish, playing energetically alone, dragging his tin plate across the gravel or trying to balance it on his head while he walked round the yard. I went into the shack to do some writing. A while later I heard Réné and Julian laughing outside. Pooh was crossing the yard, pirouetting and somersaulting, conscious that all attention was focused on him. His face, hands and feet were a ghastly chalk white, his coat a pretty gray. He looked like a warrior from some primitive tribe painted for a ritual. His face was a flat white mask with dark round eyes and a dark line along the mouth. Pooh had found a pile of white ashes with the texture of talcum powder. The smoothness and softness appealed to him—it probably reminded him of soapy foam—and he had "washed" himself thoroughly with the ashes.

Pooh playing, his face covered with ashes

During the next few days I stayed with the chimps, walking out into the valley each morning with them. The green afzalia pods had turned a brittle brown. Many of them had split open on the trees, spilling their black beans on the ground. Each bean had a bright orange cap that fitted over one end of the seed and attached it to the sides of the pod. They were attractive but as hard as small stones. I was amazed to see Tina sorting through the ashes at the base of an afzalia tree and picking up the charred-looking seeds. Once cooked, they crumbled easily between strong molars and could be scrunched into a paste that tasted like roasted peanuts. The chimps loved them, and soon Pooh and William were sorting through the ashes as avidly as Tina. I wondered if Tina had accidently discovered that the cooked seeds became crumbly enough to chew or whether this was yet another remembered legacy from her truly wild days.

As I watched the chimps I speculated whether it was a scene like this that had inspired my own ancestors to control and use fire. For over a week after the fire, large fallen trees smoldered slowly to ashes in the valley, fanned gently by chance breezes. It would be a short step from searching for the comparatively few cooked seeds to gathering the hard raw seeds and deliberately cooking them in these natural ovens that were scattered about in the valley.

A FEW DAYS AFTER the fire I received a message from Mr. Gueye, the conservator of Niokolo Koba National Park, that two Italian girls and their small chimp were waiting for me in Niokolo. Raphaella and Bobo had arrived! I raced to Niokolo, to be met first by Raphaella's friend, Barbelle, a gay, pretty girl, and then by Raphaella herself. I took to her from the start. She was dressed in a pair of dusty blue jeans and a faded green army shirt. Her chestnut hair was tousled and covered with a fine layer of dust. She had a strong, angular face and fiery dark eyes. A cigarette clenched between her white, even teeth completed my first impression of a beautiful, determined and tough young woman. Bobo strutted down the verandah toward us, very erect, and confidently climbed into my arms. He seemed to exude the same determination and self-confidence as Raphaella. There was nothing hesitant about anything he did. It was late, so we set off for camp almost immediately.

Bobo began the journey on Raphaella's lap but was soon climbing around the Land Rover. As we drove along, he came to sit on my lap. Looking solemnly into my face, he pinched my arm. I didn't respond. Still looking at me, he pinched a little harder. I glanced at him and slowly pinched him back. Bobo moved away and continued his playful scrambling in the back of the car. For two years Bobo had lived among people. As soon as he met a person, he knew whether they were afraid or not, whether he could intimidate or tease them. Just to be sure what kind I was, he had quietly tested my reactions by pinching my arm. When he moved away I felt confident that, for the time being anyway, Bobo considered me worthy of befriending. Later, I knew, he would test me again to find what limits he could reach. Only when I had earned his respect would Bobo trust me completely.

I was thrilled that Pooh would have a playmate. I imagined he and Bobo becoming firm friends. But how would William and Tina react to Bobo? It was dark when we reached camp. Pooh was sitting on the ladder, William by the tent. William ran over immediately and opened the passenger door before I could stop him. He climbed in, then stopped. He had seen Bobo. William was excited but friendly. Bobo, obviously uncertain of this comparative giant reaching out toward him, backed away. I held William while everyone climbed out of the car. Raphaella put Bobo on the ground, and William and Pooh crowded round him. Bobo clung to Raphaella's leg with one arm and with the other punched out at William and Pooh—even bit them. Neither of them showed the slightest sign of being aggressive, and slowly Bobo began to relax. His gestures became less defensive and more playful.

The introduction was carried out by the light of a hurricane lamp, so I did not immediately see Tina in the kenno above us, watching intently. Bobo was still tense, so to give him time to relax a little, I handed William and Pooh half a loaf of bread each. Tina then appeared for her share. Bobo got a smaller piece so that he would finish eating at the same time as the others. He looked tiny in comparison with Pooh and William, and I realized for the first time how much both of them had grown dring the past six months.

When the bread had been eaten, Bobo began to play with William. Pooh tried to join in, and William attacked him. Bobo started

Bobo eating a pod

toward Raphaella. William pulled him back and caused Bobo to whimper—immediately William embraced him. We all sat outside the shack till nine thirty. Tina left early and returned to her nest, having done nothing except watch intently. William and Bobo played incessantly, and Pooh took part whenever William allowed it. I was amazed at the reactions of the chimps to each other. We could not have wished for a better welcome.

Finally Raphaella tried to settle a very tired Bobo on a cushion on top of the Land Rover. William and Pooh reluctantly went back to their nests in the gully. Bobo refused to sleep away from Raphaella, and finally fell asleep in the bed with her. There was plenty of time for him to learn about nesting. He had experienced enough of his new life for one day.

As with Pooh, William, Tina and countless other baby chimps, Bobo's mother had been killed when he was still a helpless infant. Confined in a crate with other young chimps, he had been exported from his native Africa to Europe. Finally Bobo and two other traveling companions found themselves in small, cold cages on the shelves of a scruffy Italian pet shop. The other two little chimps died soon after they reached the pet shop. Miraculously Bobo survived, and it was from there that Raphaella had rescued him.

For two years he had lived happily within the security of his small human family. He had never been confined. As a result, Raphaella's house had been torn to pieces. She got tired of buying new lavatory seats, so did without. She replaced the washbasin twice. She no longer bothered cleaning up the curtains or the smudgy marks all over the walls. Her whole life had to be changed to suit Bobo. She knew that for his own sake she could keep him in Italy no longer, and began to consider taking him back to Africa and reintroducing him to wild chimpanzees. She was prepared to live in the bush with him till he was mature and capable enough to survive without her. Then she had heard what I was trying to do. The rest had followed naturally.

The next morning Pooh and William were outside the shack earlier than usual. I had hoped to get Bobo out before either of them knew he had slept there, for they would not be able to understand why this new chimp was allowed to enter and they were not. As I hurried to get Bobo out of the shack, William tried to slip past me. I bent down and took his arm to prevent him. He flew into a rage, and for the first time in his life bit my hand and then threw me to the ground with amazing ease. Réné and Julian ran to my rescue, and William attacked and bit Réné. He was furious and tremendously excited. I too was furious, and so hurt and indignant that

William had actually bitten me I had no room to feel fear. I lunged at him with such fury and determination that he began to scream and run away. I caught him, grabbed a handful of hair at the nape of his neck and bit him as hard as I could on the arm. He shrieked more loudly, used his amazing strength to pull free of my grip, and ran screaming into the valley. I lay where he had left me, shaking with exertion and spitting the hair out of my mouth. I knew, though, that I had regained William's respect. It would be a long while before he tried to overpower me again.

A quarter of an hour later William returned to camp, calm and reasonable. Bobo, meantime, had to be fed. We couldn't feed him and not feed the others, so all the chimps were fed on the edge of the gully. Then Bobo followed William and Pooh to the fig tree. Bobo came back in five minutes. William followed him closely, fascinated by this new arrival.

I decided to take William, Pooh and Bobo into the valley. William kept close to Bobo all the way down into the gully. The new sounds unnerved Bobo slightly, so he asked Raphaella to carry him. She stooped and lifted Bobo into her arms. William was angry that Bobo had been taken out of his reach, and flipped Raphaella neatly to the ground. This, of course, frightened Bobo even more, and he clung still more tenaciously to Raphaella. She tried to insist he walk beside her but he began to scream harshly, which excited William still further. Finally the problem was solved when Bobo allowed me to carry him provided Raphaella stayed close to me. William did not seem to mind if I was the one carrying the newcomer. When we reached the stream I put Bobo down, but he kept walking away from William. He was too bemused by his new environment to feel like playing. Surprisingly William resorted to whimpering for his attention. Then, when Bobo still did not respond, William attacked Pooh and ran on to drum out his frustration on the root of a mandico tree.

While we washed, William played gently with Bobo, and Pooh nested some distance away. However, the moment Bobo approached Raphaella, William flew into a temper. He flung sticks and displayed violently five times in the course of an hour. I tried to be calm and patient at first, but when William began hurling large boulders at Raphaella and Barbelle, I had to intervene with another show of strength. I had never seen William so excitable and aggressive before. I flung the rocks back at him and displayed myself. William calmed down and came over quietly, his excitement once more under control. I led the group down into the valley to find a tree where the chimps could feed and William would have something to keep him occupied.

Again Bobo began to whimper, but William gently embraced him and Bobo was reassured. To my amazement he walked beside William, holding in one hand a tuft of hair on William's shoulder. It was touching to see a small chimp walking confidently next to such a big one and to observe the concern William showed for Bobo.

We came across some dry afzalia pods that were still closed. Raphaella opened them by holding them on their sides on a rock and banging them with a second rock till the hard pods split with an explosion. The seeds inside were hard—too hard for the chimps to break with their teeth—so Raphaella ingratiated herself with William by pounding them into tiny pieces which the chimps could chew. Before long the chimps were wandering about picking up the pods and queueing up beside Raphaella for her to open them. I

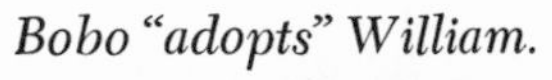

Bobo "adopts" William.

Raphaella teaches Bobo to hammer open a pod.

was astonished at how quickly Bobo was fitting into these novel surroundings. By the time we moved on, all three chimps were trying to imitate the methods Raphaella had used to make the seeds in the pods both accessible and edible. They were not successful, but they had grasped the idea, and I knew it would only be a matter of time before all of them would be opening their own pods.

At last Pooh, who had been waiting patiently while his big brother had first turn with this new playmate, was allowed to roll and tumble about with Bobo. Pooh was so gentle and considerate with the tough little stranger that it brought a lump to my throat. There would be no more lonely days for Pooh while William courted Tina. Here was the perfect companion for him.

When we reached the baobab tree William picked up a large fruit and, with some effort, managed to bite it open. Bobo watched —his face only inches from William's. William was usually very possessive about his food and I had rarely known him to share, so it came as rather a shock when I saw him break off a large piece and

Stella teaches him how to use a nest.

hand it to Bobo. I was further surprised when Bobo not only accepted the strange food but ate it with scarcely any hesitation. Bobo then approached Pooh. This time he did not even wait to be served —he reached out and confidently helped himself to almost half Pooh's fruit. Pooh looked bewildered but did not make a murmur of protest. After the chimps had eaten their fill, William lay down and rested. Bobo lay beside him. Pooh lay on a branch near them both.

In midafternoon we returned to camp. All the chimps walked, though once or twice Bobo whimpered to be picked up. I noticed William was no longer so attentive; it was Pooh who hugged and reassured Bobo this time. Bobo watched William and Pooh crouch on the bank of the stream to drink, and then did the same.

During the afternoon William left camp, perhaps to look for Tina. Pooh and Bobo were alone for the first time, and kept us all amused for over an hour with their energetic games of tag. They found an empty baobab shell and chased each other round the yard, each trying to get possession of it. Bobo became so involved in the game that he ran up the ladder to the platform without even noticing the height.

Later that evening Tina and William appeared in camp. Wary of the strangers, Tina remained in a tree on the edge of camp and merely watched. Eventually William and Tina went into the gully to nest, and we were able to give Bobo his supper and quota of milk. Pooh ate with him, and then it was time for bed again. I parked the Land Rover close to the shack and with a large cushion and some towels we made a comfortable bed for Bobo on its roof. After three or four attempts Raphaella managed to settle him in his new accommodation. The following morning I found William sprawled out on the cushion on the Land Rover and Bobo lying curled up on the ground by the door.

William was still disturbed by the new arrival. When he saw Raphaella carrying Bobo the next morning, he began to get excited again, his coat bristled and before I could stop him he was running toward her. On reaching her he bit her arm and snatched Bobo. Raphaella picked up a piece of wood and flung it at William, hitting him hard on the back. He turned round and charged at Raphaella again. Instead of running away she picked up a rock and ran to meet him. She was too angry to feel fear. William stopped in his tracks,

Bobo and Pooh at play

intimidated, and began to scream. Raphaella threw the rock down onto the ground so hard that it shattered. Still screaming, William turned to me, holding his hand out for moral support. I looked away. Never had a stranger stood up to his intimidation displays like this.

When William had almost stopped screaming, Raphaella squatted on the ground and called to him. He hurried over to her and she groomed him till he was calm. He turned and groomed Raphaella enthusiastically for a few minutes, then wandered off toward the fig tree. I heard Tina pant a greeting to him, and they both walked down into the gully.

For the rest of the afternoon Pooh and Bobo played again. This time they took Raphaella's hairbrush and, between sessions of brushing themselves and each other, ran up and down the ladder to the platform, grappling for possession of the brush, until I felt exhausted merely watching them.

24. THE BABY ELEPHANT

ONE MORNING shortly after Raphaella's arrival, a National Parks Land Rover drove into camp with a message that Mr. Dupuy, the Director of National Parks, would be visiting Simenti for a few days and had asked to see me. Réné and Julian took charge of the camp for the day while Raphaella and I shared the fifty-mile drive to Simenti.

We pulled up in the forecourt of the Simenti Hotel. A young Frenchman ran out to greet us. His name was Alain, and he was spending a few months in the park living with the guards to gain experience.

"Thank God you've arrived," he said, and gripping my arm dragged me toward the dining room. "We have a problem here we don't know how to handle," he went on. "I hope you can help."

We stopped in the middle of the dining room, and Alain looked round expectantly. "*Toto, où es tu?*" he called, and from behind one of the tables emerged the smallest baby elephant I'd ever seen. It stood 28 inches high and was approximately 10–14 days old. It walked quickly toward Alain and began to search all over him for a nipple to suck. Raff and I were momentarily stunned, then walked slowly toward the small creature and touched its bristly little body.

Alain squatted down and let the baby elephant suck his fingers. He looked up at us anxiously and asked: "Do you think we can save him? His mother was killed by poachers about twelve kilometers from here. He wandered into the hotel by himself two days ago. He has adopted me, he even sleeps with me, but tomorrow we are going on a week's patrol of the river and no one knows how to take care of him. I hope you can take him. You have looked after young animals before, haven't you?"

The elephant interrupted Alain's string of words with a noisy jet of liquid diarrhea. I felt what little hope I'd had of being able to rear such a young elephant diminish even further. From my time at Woburn I knew that in one so young and small diarrhea was an almost certain killer, and I had read that such young elephants were almost impossible to hand-raise anyway.

My thoughts flew back to the orphans I'd raised in The Gambia, and to the diet that the three baby elephants had been started on at Woburn: a gruel of cooked maize meal, milk, glucose and calcium. These had been much older than this tiny infant, but their diet was the only one I knew and there was no one else around who had even this much experience. Yet the only ingredient of that diet to be found at Simenti was milk. I decided to substitute cooked rice for the maize meal and honey for the glucose and arrange the rations as best I could. In camp we had plenty of calcium.

I asked Alain not to feed the elephant any milk for the rest of the day but to give it warm water with honey and a trace of salt. This was an attempt to clear its stomach, giving it a rest before we started the elephant on a new diet.

At midday I sent a cable to Daddy and Nigel by radio, requesting some of the lambing teats I knew we had in the house and included a list of other requirements for the young elephant. I also asked them to phone or cable zoos in England that had successfully raised elephants to inquire about suitable diets. In the meantime we would have to make do with what was available.

Mr. Dupuy and Mr. Gueye arrived soon afterward. They had a short discussion, and finally Mr. Dupuy turned to me and asked if we could take the baby elephant to camp and attempt to rear it. The Parks would supply the milk and whatever else was necessary. From the moment we'd seen it both Raphaella and I had badly wanted to look after this endearing baby elephant, but I wasn't sure how feasible it was going to be to have it in camp. Before I could think of any reasonable objection, however, the matter seemed settled. We were the best-equipped people in the park to take over the task, and it was ours.

Despite my fears, I was aware of my terrific excitement at the prospect of looking after the baby elephant. Suddenly the little creature was a member of the family, and it was vitally important to

The baby elephant asleep

me that it should survive and grow up. This was another adopted orphan—another young life I held myself responsible for.

Raphaella and I set about equipping the Land Rover for the ride home. The guards gave us a straw mattress; we laid it in the back and cut a pile of springy green leaves to put on top of it. We collected some old curtains and one or two sacks in case it got drafty, filled a plastic jerry can with boiled water and commandeered the hotel's supply of honey and several plastic water bottles. Alain made teats from the fingers cut off a pair of rubber gloves. Then he brought the elephant to the car, and we gave it a final drink of water and honey.

The elephant was still frequently passing a waterlike diarrhea. As I watched, something struck me as rather odd. Alain had said it was a male elephant, but suddenly I had doubts. To prove his point Alain pointed to a protuberance under the belly. I looked more carefully and saw that what we had all mistaken for evidence of the elephant's male sex was in fact the remains of the umbilical cord, half obscured by a flap of skin. The cord seemed wet and slightly infected, and was caked with a slimy yellow excretion. We cleaned it immediately with cotton wool and some antiseptic lotion, then sprinkled it liberally with antiseptic powder.

Finally at four o'clock all seemed ready. I bent down, lifted the little creature into the back of the Land Rover with surprising ease, and climbed in after her. She was very nervous and began immediately to try to find a way out. When I held her back she rumbled, then gave a loud scream of distress.

The baby elephant was strong for her size, and it was quite a job holding on to her. I spoke to her constantly, trying to soothe her, and Raphaella drove at about 5 mph in order to give her as smooth a ride as possible. After struggling to keep her balance for half an hour, she lay down with her head cushioned on my thighs. We had to crawl along, easing the Land Rover in and out of every hole so as to avoid jolting her. She was young, distressed, sick and a very delicate passenger indeed. Furthermore, she was extremely restless and was up and down like a yoyo. I got covered in diarrhea and the honey water—which we gave her once every hour—and was thoroughly trampled upon. It had taken us three hours to reach Simenti that morning, but it took over nine hours to get back to camp.

Réné and Julian couldn't believe their eyes when we arrived; in fact, at first they seemed almost afraid of the baby elephant. We lifted her down from the Land Rover, and she followed us into the shack, nudging us for a drink. I put some rice on to cook, and half an hour later gave her half a liter of rice, milk, honey and calcium gruel. Sucking on an improvised teat, she drank it all down—her eyes half closed with satisfaction. She seemed perfectly at home in the shack and, when she had finished her meal, climbed up on the double camp bed beside Raphaella and lay down to sleep. We looked at one another, smiled and shrugged. I made up another feed, which I put in the fridge, and climbed into bed with Raphaella and the baby elephant.

Fifteen minutes later, just as I was dozing off to sleep, the baby elephant stirred, got up, walked over my stomach and stepped down onto the floor. Then she began to sniff my face with her wet, rubbery little trunk. Finally she knelt down, rolled her trunk back onto her forehead and began to slobber all over my neck and shoulders with her tongue, searching for a nipple. I sat up and allowed her to

suck on my fingers. I looked at my watch. It was barely twenty minutes since her last feed. I wasn't sure how much an elephant of her age could safely drink at one time; I didn't want her to go hungry, yet on the other hand I knew it was dangerous to overfeed her—especially when she already had diarrhea. So I tried to soothe her as best I could.

It took her about ten minutes to realize that the fingers I kept giving her to suck produced no nourishment, and then she began to butt me hard with her forehead and push me around. Soon she began to rumble, and finally she screamed and butted me again. I was amused at all this determination and willfulness in one so young. I sat through half an hour of buffeting and then warmed up the feed I had in the fridge. The baby elephant sucked strongly till there was only half an inch of gruel left in the bottle, then, apparently satisfied, climbed back onto the bed, flopped down on Raphaella, who merely stirred and wriggled into a better position, and fell into a deep sleep. I cooked some more rice and made up two more bottles, which I stacked in the fridge. I was sticky with saliva and my face and arms were sore from the abrasive contact with the elephant's bristly hide.

It had taken me almost an hour to prepare the two new feeds, and as I edged my way into our now rather soiled bed, the baby elephant woke up. Refreshed by her doze, she seemed desperately hungry and once again began to suck and nuzzle any part of me she could find. I almost woke up Raphaella and told her it was her turn to baby-sit, but she was so soundly asleep that I decided against it. I tried to ignore the elephant, hoping she would go to sleep again, but instead she trundled round the room knocking everything over, then came back to me in a slight panic at having lost contact for a few minutes and began again the eternal nudging and slobbering for food. Finally, out of sheer desperation, I had to surrender and give the baby elephant another feed. After that we both fell asleep for one glorious hour, until the whole process resumed. It was almost light by this time. I felt extremely tired. My mothering instincts and patience were wearing noticeably thin, but I got up and gave the

Stella and the elephant

baby yet another feed. Raphaella woke up soon afterward and took over, so that finally I was able to have a few hours of uninterrupted sleep.

Next morning, after a cup of strong coffee, we led the baby elephant outside. The chimps had seen her in the shack and were very curious now that she was outside and within reach, though they were also rather afraid. Finally Bobo came forward, took hold of her tail and sniffed it. I watched carefully to make sure that he didn't pull or bite it, but when the baby elephant turned around, Bobo let go and backed away. Only Pooh, in his pleasure at seeing us, ignored the baby elephant and came to sit in my lap. He kept pushing the wet trunk away each time it blew or snuffled at him, but did not seem unduly alarmed.

When I saw William entering camp I slipped the starter's pistol out of my pocket and held it hidden in my hand. I spoke to him, introducing the baby elephant, and trying to act as if everything was perfectly normal. He approached cautiously, and I allowed him to sniff the elephant and to look behind her ears and under her tail. Then he gripped her trunk and probably squeezed it, because the baby elephant began to shake her head. I took William's hand and, chatting brightly, firmly lifted it away. Without being in the least aggressive I was making sure that he understood the elephant was as much a part of me as my bag and my camera and that, although he could touch her, she was mine. Surprisingly, William accepted all this very casually, and after another brief inspection he left her alone.

The baby elephant trailed Raphaella and me like a shadow. If we weren't together the elephant seemed in a dilemma as to which one of us to follow. Finally, so long as she was close to one of us, she appeared content. She still had awful diarrhea. Raff had brought a box of medicine for the chimps that included several different types of liquids and tablets to combat diarrhea in human infants. She spent half an hour going through the box, and at last came up with a medicine which we both thought would be suitable for an elephant.

We took turns looking after the elephant so that at least one of us got some sleep every other night. The days went by—each minute seemed devoted to the baby elephant. Soon we were taking her for short walks with the chimps to visit some fruiting tabbo trees close to camp, and in the evening she would come down to the stream with us. Despite the intensive care she demanded, we found her a most endearing little creature. She was an infant, and dependent on us, yet had strength of character and was often willful.

A week went by and Nigel arrived with all the things I'd asked for in the cable. He had no diets for baby elephants, but he brought with him a book on elephants by Iain Douglas Hamilton and a chemical breakdown of elephant milk. Without the use of a laboratory, though, it would be impossible to simulate natural elephant milk, so we continued with the diet we had. Now that we had glass bottles and proper teats, it was much easier for Réné and Julian to keep things sterile. The fridge was by this time entirely devoted to baby-elephant feeding equipment. From the elephant book, we guessed that the baby elephant was approximately two weeks old—very young indeed.

Nigel also brought something less welcome—a telegram which meant Raphaella had to leave with him to return to Italy. She was very distressed about having to abandon me, but I assured her I would manage somehow and that she shouldn't feel bad about going. The baby elephant and I helped her to pack that evening, for Nigel was going back to Gambia the following morning. Nigel promised he would explain the situation to Daddy and said he was sure he could come to camp for a couple of weeks to help me out.

After Raphaella and Nigel had left, I walked back into the shack with the baby elephant. Very rarely in my life had I felt quite so lonely and depressed. There was little time for brooding, though, and I quickly got my boots on and took the elephant and chimps out to the tabbo trees. I could not go too far from camp, for every two hours I had to go back to the shack to give the baby elephant a drink. Her stools seemed to be improving slowly. Instead of ejecting a stream of water, she now passed something of about the consistency of pudding. I began to hope that the impossible had happened and that she was going to live after all.

The first two nights I slept little, and as the baby elephant would not remain with Julian, there was little chance of rest during the day either. Nor could I completely rely on Julian to measure out the ingredients for a feed; it seemed extremely difficult to impress upon him that every quantity had to be exact. Rather than risk upsetting the elephant's stomach again, I made all the feeds myself. The third night I felt so desperate for sleep that I strung up a hammock in the room. A feed every two hours was still essential if she was to survive, but I could count on her to wake me up well before that time. Safely up in the air, she would still feel I was near her yet she would not be able to trample all over me or soil my bedclothes each time she got up to stretch her legs.

After Raphaella had been gone a week I was relieved to find that I was coping. Each morning and afternoon I'd lead the chimps a short way into the valley or along the plateau in search of food, and the baby elephant would walk along in the midst of them. William soon got bored with these short walks and spent most of the time with Tina somewhere out in the valley. Pooh became attached to the elephant and frequently walked with an arm around her neck, while Bobo occasionally patted her bottom or walked upright behind her like a midget *mahout*. Except for having to return to the shack for her feeds, she was little trouble during the day.

One afternoon while I sat watching the chimps feeding, the elephant stood with her head over my shoulder, leaning her weight against my back and dozing. Her long black eyelashes kept quivering against my cheek. I put up my hand to rub her face fondly, and she immediately guided it with her trunk to her mouth and began to suck my fingers. As she sucked I became aware of something sharp on her gum, and when I looked I saw that she was cutting a tooth. Her gum was red and a tiny piece of the tooth was visible. I pressed my face against hers and congratulated her, and she wiggled her trunk playfully in my hair. I shall be so relieved, I thought, when you get more of those teeth and start eating solid food, when you get past this delicate stage. We can all go for long walks together. Perhaps you might even meet other elephants and make friends. Camp would be the perfect home for a young elephant to grow up in.

That evening I sat out on the anthill in the yard till all the chimps had nested, then went into the shack for supper. The baby elephant got her bottle first, but kept trying to find out what I was eating and put her trunk into everything on the table. When I pushed it off she walked round the room confidently, clumsily sniffing here and there and knocking what few things remained on the shelves onto the floor. She waved a towel around, and wandered back to the table and mouthed my thigh till I had finished eating.

I strung up the hammock, and at nine thirty gave her a feed and went to bed. She climbed onto the camp bed and settled down too. I turned off the flashlight and fell into a sound sleep.

I woke with a start. Everything was so quiet. I switched on the flashlight and looked at my watch—it was quarter to four. Oh God, the feeds. The baby elephant, where was she? "Baby Elephant!" I called, and shone my light frantically round the room. The camp bed was empty, but just below the hammock I saw the elephant lying on her side. I leaped from the hammock and crouched beside her. There was no response, she was cold and unconscious.

I yelled for Julian and a few moments later he rushed in.

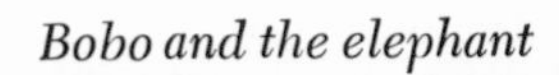

Bobo and the elephant

"What's the matter?" he asked breathlessly.

"Light a fire, a big fire, and warm up as much water as you can. Then fill all the empty bottles you can find," I cried.

I covered the little elephant with blankets to try to warm her. My insides were a tangle of guilt. I had overslept, she had missed three feeds and I was responsible for her collapse. I lifted her head, praying for some signs of life. She had been so well at suppertime; it seemed impossible she could be so lifeless now. Her breathing was slow and labored, as if each breath was held for seconds before it was exhaled. Torn with anguish and feeling helpless, I cradled her head. She stopped breathing, but unable to accept what that meant, I began to pump her chest and lift her front leg alternately. After I had forced air in and out of her several times, she began to breathe again alone. We packed hot water bottles—wrapped in towels, shirts, anything we could find—all round her, and I rubbed her body all over with my hands in an attempt to keep the circulation going.

If only she could regain consciousness and I could feed her, give her some nourishment, I felt there might be hope. My elephant couldn't die, just couldn't die. I loved her too much. I needed her.

In desperation I gave her a tablespoon of brandy, making sure the liquid went down her throat and not into her lungs. Whether it was the result of the hot water bottles, the brandy or an iron will to live, about a quarter of an hour later she opened her eyes and seemed to focus. Her trunk wriggled freely. I prayed that she would wake up enough to drink. Julian ran to the kitchen for the bottle and gave it to me. I held the teat on her tongue, but she was still much too groggy to drink. I put the bottle down and continued to talk to her and rub her. Her eyes closed, her trunk went limp and she relapsed into unconsciousness. For an hour we rubbed her and kept her warm, then her breathing began to falter again, and when it stopped there was nothing I could do to make it start. I felt for a pulse and found nothing but stillness. My baby elephant was dead.

I took the pillow from the hammock and cushioned her head, then lay on the bed fighting for some equilibrium of mind. I felt spent—too empty and disappointed to cry, only enough feeling left to ache. Julian made an attempt to console me.

"It is the will of God, you must accept, it is the will of God, Stella."

"Why then did God let her walk twelve kilometers to Simenti when her mother was slaughtered? Why did God allow me to take her to camp? Why did he let her live for over two weeks with diarrhea, let her get better, let me love her, let us all hope, then snuff her life out like a candle? If that's God's will, I hate him!"

Julian shook his head sadly and walked back into the kitchen.

Before the chimps were up I drove the Land Rover into camp, placed a mattress in the back, carried the baby elephant out and laid her on it, then covered her with a blanket. Later that day I had to take the remains to Niokolo and report her death.

I reached there late that afternoon and drove straight to the office. The veterinary officer was waiting with Mr. Gueye, and almost as soon as the vehicle stopped, a group of guards unloaded the elephant to take it to the dissection room. It was all I could do not to yell "Be gentle with her!" Mr. Gueye asked me all the details of her death, and said it was a pity. "I tried, Mr. Gueye, believe me, I tried," was all I could answer. The sight of the knives being laid out made me want to vomit.

25. YULA AND CAMERON

SOON THE TIME CAME for my journey to England. Almost the first thing I did after arriving in London was go to London Zoo and meet Dr. Brambell. This time I was allowed to do more than merely watch Yula and Cameron through the bars of their cage. The chimps were now each five years old. When I entered their cage Cameron grew excited and played boisterously; Yula climbed into my arms and sat there, quietly grooming my face and clothes with slow determination. Occasionally she'd attempt to imitate her brother and make a few play gestures, but most of the time she preferred to sit and watch. Toward the end of my visit she was beginning to relax and move around a little more, playing with her keeper and me, but never with the same zest as Cameron.

Each time I meet new chimps it is an exciting experience, but knowing that Yula and Cameron were soon to be part of the family in camp made me feel even more exhilarated than usual. Yula and Cameron would be spending about six months in Abuko to become acclimatized. In anticipation of their arrival, Daddy had pulled down the old chimp enclosure at Abuko and built a new one, slightly smaller but strong and durable.

As I could not be in camp and at Abuko at the same time, I had clearly reached the stage where I needed an assistant. Hugo knew an American graduate student, Charlene Coalglazier, who was anxious to come and help. She flew to London from America in January, a few days after me. She was a cheerful, strong-looking girl with long dark-brown hair. She had not had the chance to gain much practical experience with chimps, but what she lacked in experience she made up for in enthusiasm. Charlene came to London Zoo with me on two occasions to meet Yula and Cameron, and played with them both confidently.

Charlene was to be in charge of Yula and Cameron while they stayed at Abuko. Each day she would take them out into the reserve, introducing them slowly to their new surroundings, to climbing and to the new foods that the forest could provide. Yula and Cameron had spent all their lives in the zoo. I suspected that they would already be set in their ways and therefore less adaptable than their predecessors. Teaching them a new way of life would take a lot of time, patience and understanding.

The journey back to The Gambia was mercifully uneventful. A group of men were waiting for us at Abuko, and the crate was unloaded and carried carefully into the enclosure. Cameron and Yula peered out of the wire grille. They looked bewildered and afraid. Yula clung desperately to the only familiar figure in this hot, new world—her brother Cameron.

The crate was placed in the middle of the enclosure. I unlocked the door and opened it. There was no desperate rush for freedom. Yula and Cameron sat still, looking out suspiciously. For several minutes neither moved; then Cameron advanced cautiously toward the entrance of the crate and peered out. He took one more short step and his knuckles came into contact with an unfamiliar surface —warm, dry, sandy soil. He pulled his hand back into the crate, almost as if he had received an electric shock from this new texture. He forgot about looking around; all his concentration seemed directed at the ground.

He touched the soil again hesitantly, then put some weight on his hand. His knuckles sank in an inch before the ground seemed firm. Cameron was fascinated. He sat down, scooped up a handful of the warm sand, then let it trickle slowly through his fingers. He was, I am sure, merely satisfying his curiosity, but I could not help being amazed at the rightness of his gesture in the circumstances. How many men after long exile have greeted their homeland by embracing its soil. How many men have expressed their love of country with the same gesture Cameron had just used.

Carefully Cameron emerged. He walked clear of the crate, then with gathering confidence moved faster till he was almost running the length of the enclosure, touching the feeding table, the bench, the sleeping quarters, the struts to the climbing apparatus. He was so involved in his exploration that for about five minutes he ignored everyone present, including his sister, who rocked nervously to and fro at the back of the crate. I called to her, offering her comfort, inviting her to come out. At first she seemed not to hear, but then, as Cameron began taking notice of the group of men outside the enclosure, she came forward slowly and looked about to see what was causing the hum of conversation and the occasional laughter.

Yula also was suspicious of the ground, but after testing it briefly she hurried into my open arms and sat on my lap. I embraced her and tried to make her feel as secure as I could in what must have been a traumatic situation for them both: a rebirth, a total change. After sitting in my lap and staring around her for a few moments, she climbed down and hurried to Cameron. They embraced each other, then together made a tour of the enclosure.

In the sleeping quarters the two hammocks were filled with wood-wool, which was the bedding the two chimps had been accustomed to in the zoo. To my dismay they pulled the wool out of the elevated sleeping quarters and carried it to the ground. They arranged it into a vague nest shape around themselves and prepared to sleep on the ground close to one another. I tried once to put the wool back into the sleeping quarters, but they pulled it all out again. I decided they had experienced enough changes in the last twenty-four hours and allowed them to choose their own sleeping sites.

For the first three days I spent as much time as I could in the enclosure with the chimps. They adjusted quickly to the space and strangeness and soon adopted me as an acceptable foster parent. Yula especially seemed to rely on me, and enjoyed the affection and attention I gave her. They both obviously found it hot at midday and became sleepy and lethargic, but during the morning and evening they often played with each other. Neither of them played on the climbing apparatus, though, and both refused to use the enclosed sleeping quarters. Perhaps it reminded them too much of the crate; perhaps they were already beginning to value space.

On the fourth day I decided to take them for their first walk. I carried Yula out of the enclosure. Cameron followed us confidently

—far more confidently than I'd expected. Once in the forest, I sat down. Yula scrambled off my lap and hurried to her brother, and they both set off down the narrow path as if they knew the area intimately. I trailed quietly. Cameron left the path in an open patch and walked into the vegetation. Yula hesitated, then followed him. I began to feel uneasy that I had overestimated their reliance on me. If they began to explore independently they might easily get lost. I had lost sight of the chimps for approximately one minute when Yula began to scream. I ran over. She was screaming because she couldn't find me—she was afraid to be alone in this strange new world—and as soon as she saw me, she hurried over and climbed into the security of my arms.

Cameron was much more independent—worryingly so—but I knew that as long as Yula kept close to me, Cameron would too and I could supervise their first introduction to an African forest. Neither of them showed the slightest inclination to climb the trees or the lianas, but I found it encouraging that on his first walk Cameron consented to taste a wild fruit and liked it. The green vervet monkeys had thrown many ripe fruits to the ground, and Cameron began to pick them up and eat them. Yula was too awed to feel hungry, and didn't even seem curious about the fact that Cameron was eating.

I kept the two chimps out all morning, and they both enjoyed the walk. Cameron especially raced up and down the path in exuberant high spirits. I was anxious lest it might be difficult to persuade him to re-enter the enclosure, so Abduli had placed a small feast and two tempting mugs of fruit juice on the feeding table. I carried Yula in and gave her her drink; Cameron followed us in almost as eagerly as he had followed us out, and drank his juice. Then both of them began to eat. They still refused the papaw and other local fruits, and stubbornly ate only the fruit with which they were familiar.

As soon as Charlene arrived she began to acquaint herself with the two chimps and learn how to take care of them. For two weeks she took the chimps out for walks together, till they became familiar with her and she with them.

WITH CHARLENE IN CONTROL I was free to go back to Niokolo, so 14 days after my return from England, I left Abuko. In all, I'd been gone for 6 weeks. The chimps were late going to bed the evening of my arrival at camp, but finally William wandered down into the gully and climbed into an old nest. Pooh went to sleep in a nest made of vines at the back of the shack, and Bobo climbed to the platform alone and arranged his pile of leaves in a nest shape before settling down. He no longer needed his cushion, and Nigel, who had been in charge during my absence, had gradually weaned him away from his towel blanket.

I talked with Nigel for a long time after supper, catching up on all that had taken place in camp. Shortly after I'd left, Bobo disappeared for thirty-six hours. Almost beside himself with worry, Nigel had searched the valley several times. When Tina returned alone, Nigel's anxiety increased. Finally Julian had found Bobo half a mile down the stream sitting in a tree and had run to get Nigel. Bobo was extremely relieved to see Nigel and had hurried into his arms. He had a cut above his eye which was swollen and bruised and had split the nail on one of his big toes, but was otherwise unhurt. Nigel suspected that he had fallen out of a tree and injured his head, as he had not responded to the calls Nigel, Julian and Réné had given while out searching for him. Bobo was carried back to camp and nursed for a day and had since shown no ill effects of his adventure.

Tina appeared next morning from the gully. I had been waiting for her to turn up and panted my enthusiastic greeting. She climbed down, hurried toward me, gripped my chin firmly, laid her open mouth on my neck and panted back at me. As usual she looked extremely fit, big and solid with a fine gloss to her coat. Bobo came running toward us and flung himself at Tina. She turned to him at once, took his small arm in a capable grip and began to groom him. Bobo had certainly become very familiar with Tina, and she was prepared to indulge and mother him to a surprising degree. I was reminded of the way that she had adopted Happy when she was scarcely more than a child herself, and wondered if I would ever have the thrill of seeing Tina with a baby of her own.

From the days of Abuko, Pooh had shown a keen interest in my note-taking and, if given the opportunity, would scribble for long

A kiss from Tina

periods on a page of my book. The older he became, the longer was his attention span, and his heavy babyish scribbles began to take on a different form. When in the mood he would whimper and beg for my pen and notebook, and if for some reason I had to refuse he would work himself up into quite a temper. If I handed him pen and book he quieted immediately. Turning to a clean page, he would assume an expression of intense concentration and begin to make the lip-smacking sounds he used during a grooming session.

Over the years I watched his scrawling develop into what now closely resembles shorthand. There is no doubt in my mind that Pooh is not content with just putting pen to paper and making long, easy lines haphazardly across a page. He wanted to perfect his imitation of my note-taking. When given a pen now, he sometimes holds it as I do, but usually employs the precision grip typical of chimps. With a much-practiced control of his pen, and almost painful concentration, he makes small lines and dots above the printed lines of the notebook, working from left to right. Occasionally he also follows the red margin line, making his own small hieroglyphics from top to bottom. Sometimes his designs take another form, filling one corner of a page with a concentrated mass of lines and dots, usually the top or the bottom right-hand side of the page.

Pooh "writing" and (below) his "notes"

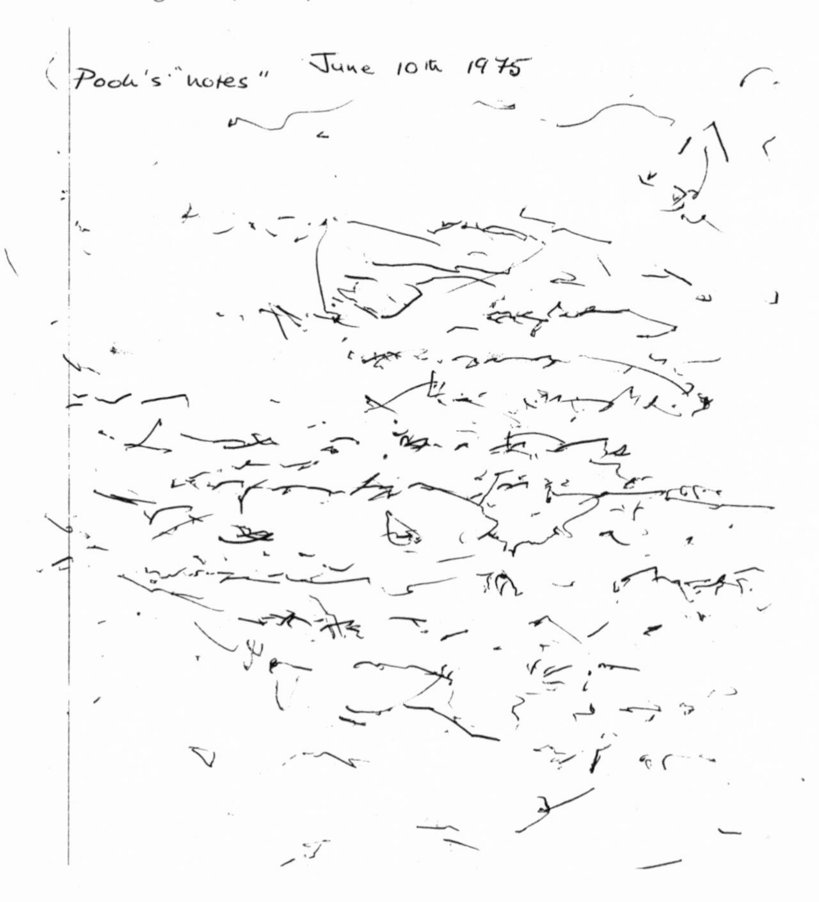

Pooh has never been taught or even actively encouraged to write; for him getting access to pen and paper is a great treat in itself. In fact, on occasions I've been positively discouraging, for it can be inconvenient to hand over notebook and pen when they are in use. I have never showed him the human way to hold a pen, and it came as a great surprise the first time I saw him use it, but it seemed that as he came nearer to producing something very similar to writing, he also wanted to improve his version of my pen grip. I often wondered whether, if Pooh had received proper tuition, he might have become adept at expressing himself in some sort of written language.

One of the foods in season at that period was a round fruit the size of a Ping-Pong ball which grew on an exceptionally thorny bush. When the fruit was ripe its casing was a rich chocolate brown and extremely hard. Tina was the only one among us who could open the fruit with her teeth. William, Pooh and Bobo learned to use rocks as hammers to shatter the rind. The flesh of the fruit was almost black and tasted like caramel. It was studded with small, bitter seeds. The chimps loved the black flesh. They would scoop it out of the shell with a forefinger and suck it till all had dissolved except the seeds, which they then spat out.

The caramel fruits were so well protected by the thorns on the parent plant that the chimps could not climb to pick them. Fortunately, when they were ripe the slightest breeze detached them from the branches, so there were usually some lying on the ground beneath each bush. William, however, was not content with eating only the fruit that had fallen, especially when he could see more tempting objects still attached to the thorny branches above him. With no human help, he devised a method of obtaining the ripe fruit. He seemed to find it easy to trace one branch through the tangle of others to its origin on the trunk of the bush. He would carefully pick away the thorns to clear an area which he could grip comfortably, then shake the branch vigorously till the fruits fell to the ground.

Pooh and Bobo soon understood that the fruits William shook down were his and his alone. He was quick to punish either of them

Teaching Bobo to open caramel-like fruits

if they attempted to pick them up while he was still shaking the branch. Once William had collected as many as he could carry, he would walk away a short distance to feed. Only then could Pooh and Bobo share what fruit remained. However, they both mastered William's method of obtaining the previously inaccessible fruits and, with the added knowledge of using a rock to hammer the fruit open, another source of food was made easily available to the chimps. Bobo was still very much a new boy, but he had a healthy appetite and, being so adaptable, willingly tried all the leaves and fruit that Tina, Pooh and William ate. I kept wishing that Raphaella could see her Bobo's progress. He was a chimp to be proud of.

AFTER 4 MONTHS IN ABUKO Yula and Cameron were ready to continue their journey. Nigel came up again to replace me in camp and I drove back to Abuko. It took us two tiresome days to persuade them to enter the crate we had prepared for them. I could have carried Yula to the Land Rover and simply placed her in the back, but Cameron, I felt sure, would never have cooperated. Finally Yula and Cameron got over their fears, entered the crate and were seated safely in the back of the car.

It was a cool but dusty ride. The chimps slept little and we stopped frequently to give them a drink. By dawn we were well into Senegal. We traveled until the sun rose, then pulled off the road and parked beneath a huge mango tree to rest. My eyes felt sore with the dust and the effort of keeping them open. We hoped that during the day Yula and Cameron would catch up on their sleep. They seemed comfortable and relaxed once the car had stopped, and they both drank and ate well. Charlene and I took turns sleeping during the day. As soon as the sun lost some of its fierceness, we moved on. By eleven the following morning we had reached the turnoff to Mount Asserik. Camp lay only two miles ahead. Nevertheless we waited till four o'clock that afternoon in some deep shade to give Yula and Cameron a chance to rest and to prepare them for the excitement of seeing the other chimps in camp. By this time they seemed to have become quite accustomed to the Land Rover and the strangeness of traveling. They ate well and drank quantities of fruit juice—Yula even began to invite Cameron to play.

I drove the last two miles to camp with a mixture of excitement and apprehension. Yula and Cameron were much older than Bobo had been on his arrival, and I was not at all sure that William would accept them so easily or that they would tolerate the other chimps. I had a nagging dread of Yula and Cameron becoming afraid and running away into the bush before I could really establish them in camp—with no experience of the wild, they would almost certainly perish. I was also worried about how Charlene would be treated. Pooh and Bobo, I knew, would accept her immediately, but I was not certain that she had as yet acquired enough experience with chimps to handle William, who was becoming more and more self-assertive in the camp. There was only one way of finding out, though, and that was to try.

26. INTEGRATION

EVERYONE WAS IN CAMP to greet us. I received a terrific welcome from William, but before I could introduce him to Charlene he had spotted the two new chimps in the back of the car. He leaped onto the front seat and grinned at them excitedly. Pooh and Bobo crept in beside William. William was perfectly sociable until Cameron rudely banged the mesh in front of his face in an aggressive manner. William began to bang back, till the Land Rover rocked and exploded with the displays the two males were acting out on either side of the mesh. There were, of course, lulls in the racket when both parties merely watched each other, but for the sake of the Land Rover as much as anyone, I felt it was important to move Yula and Cameron out as quickly as possible.

I did not dare simply to let them out. I wanted the chimps to get accustomed to each other first through a safe barrier and to give Yula and Cameron a chance to adapt to the camp. That evening the trailer cage that Claude had built exactly one year before to transport Tina was put into use again. We pushed it over to the Land Rover and positioned it so that Yula and Cameron could enter it the moment the grille of their crate was pulled open. I expected them to be so fed up with the Land Rover by this stage that they would eagerly walk into the more spacious trailer, but they were wary of everything—the trailer and other chimps included. The Land Rover had proved secure against William's buffeting, and they intended to remain there. It took us almost twenty-four hours to persuade them to take the plunge. Once they were safely in the trailer, we pushed it into the shade and secured it by piling rocks around each wheel.

Yula and Cameron were easily visible in the trailer, as the entire

William greets the new arrivals.

upper half of it was constructed of diamond steel mesh, so there were more possibilities for interaction among the chimps than there had been in the Land Rover. Yula and William soon seemed fairly friendly, and during the hiatuses between William's violent displays they sometimes entwined fingers through the mesh. But Cameron would always spoil things by poking William or pulling out tufts of his hair, which would cause him to fly into a fearful rage. I was afraid William would turn the trailer over in his efforts to intimidate Cameron, for he used every ounce of his considerable strength to bounce and rock it. He became increasingly frustrated that he could not actually reach Cameron to teach him some manners and show him who was boss of the camp.

Charlene, Nigel and I took turns watching and recording what happened. William had been so engrossed in the two new chimps that he had scarcely paid Charlene any attention, and we were all lulled into a false sense of security. Instead of Nigel or I accompanying her whenever she went outside, she began to venture confidently out of the shack alone.

Yula and Cameron were in the trailer for four days. During the hot, lethargic hours of the fifth day we intended to release them, hoping that the heat would drain away some of the energy that might otherwise be spent on fighting. The morning before the release was planned Charlene went out of the shack to record one of William's particularly violent workouts on top of the trailer. I was sitting on the Land Rover roof rack watching William when I saw him leap from the trailer, grab a hefty bamboo pole and charge straight for Charlene. He seemed enormous as he flashed by, every hair on his body stiffly erect, his powerful shoulders hunched and a grim, tight-lipped look of fury on his face. With scarcely any apparent movement of his arm, the bamboo pole left his hand, became a blur in the air and landed with a vibrating twang just behind the American girl. I only had time to shout his name before he leaped at Charlene. She gave a bloodcurdling scream as William's weight threw her against the wall of the shack. A shot from the starter's pistol stung the air, and William charged on past the shack and down into the gully. Charlene stood slumped against the wall, both hands clasped tightly to the right side of her head. We helped her into the safety of the shack. Her face was a pasty white and she shook all over as I made her sit down on the edge of the bed.

"Why did he do that?" she kept asking. "Why does he hate me so much? Why did he attack me like that? Why, Stella? What did I do? Oh God, my face!"

Blood began to glisten between her tensed fingers and to trickle down the back of her hand onto her shoulder. Nigel was already busy lighting the fire to heat up some water. Talking as soothingly as I could, I carefully lifted Charlene's hands away from her head. Blood ran freely from her ear, contrasting vividly with the pallor of her cheeks. Charlene looked at her hands, then stared at me, her eyes wide with horror.

"Dear God, what did he do?"

"It's okay, Charlene. He's bitten you, but the blood's making it all look far worse than it is. Hang on; when there's some warm water I'll clean it up and you can have a look in the mirror."

Once over the initial shock, she became quite cheerful. "God-damn it, he's a big bastard!" she drawled, and began to laugh. She lay on her side while I bathed her ear. Once it was clean it was easier to assess the damage. There were a couple of bruised teeth marks behind her ear and one canine puncture, but the bleeding was coming from halfway up her ear, where a quarter-inch piece was missing. Trying to keep my own voice steady, I told her that she was minus a piece of her ear. To my surprise, she laughed again.

"Think my ear's still out there in the dirt, or d'you think he swallowed it?" she asked.

As I bandaged her head, I tried to assure her that William did not hate her and to explain why I thought he'd attacked her. William was nine years old—he was adolescent, full of struggling emotions and just beginning to realize his own strength and power and feel the elation of independence. He was conquering his childhood fears one by one, slowly doing more of what he pleased when he pleased. He was dominant over the other chimps and knew that he could successfully intimidate all but a handful of humans. It was Cameron's bravery in the security of his cage that angered William. Even after one of his most furious displays of strength, Cameron had remained unimpressed and dared to poke him spitefully when he flopped down exhausted on top of the trailer. Finally William had become so frustrated at not being able to reach the insolent Cameron that he had redirected his aggression at Charlene. She was, until she proved otherwise, an ideal subject on which to vent his anger safely. Charlene was now thoroughly afraid of William, and no matter how hard she tried to hide her fear, William would sense it almost immediately and enjoy trying out his newfound power. We knew that it would be almost impossible for her to remain in camp with him. Also, Charlene's ear needed proper medical attention, for the risk of infection was high in the primitive conditions of the camp.

Charlene was extremely brave about the whole episode. I packed her clothes for her, and though neither of us mentioned it, we both knew that she would not be returning. For Charlene, it meant an abrupt end to eagerly prepared plans and adventurous dreams. For me, it meant a solitary existence again. I had been looking forward to Charlene's help and companionship and felt a deep sense of loneliness at the thought of continuing, perhaps for years, without anyone to share my part of the project. William was successfully isolating me, making camp into an island. He had shown that he would no longer complacently accept strangers. If anyone was to succeed Charlene, he or she would have to be an exceptional person to win William's respect and affection.

NIGEL DROVE CHARLENE to the hospital. Once they had left I felt thoroughly depressed. What was to happen next? I kept hoping that Yula and Cameron knew me well enough to have confidence in me when they were released. I could not hold them in the trailer much longer, and I was beginning to wonder if there was, in fact, anything to be gained by keeping them separated from William. Their inaccessibility seemed to make him angrier by the day. I decided that after lunch I would let them out.

Locked in one of the trunks I had a small bottle of Sernylan, a drug I kept in case I ever had to tranquilize any of the chimps. I resolved to tranquilize William before releasing Yula and Cameron. I prepared only a fraction of the required dose, though, for I did not want him to fall sound asleep only to attack Yula and Cameron when he woke up. I wanted him to be aware of what was happening but to feel too lethargic to react violently. I measured the dose with a syringe, then added it to a cup of orange juice. William came to me the moment I called, and drank the contents of the cup without a second's hesitation. I went back into the shack and loaded the starter's pistol with the special blank cartridges.

I asked Réné and Julian to stay near the group of chimps but not to intervene unless I gestured. While I was talking, there was a pant-hoot in the valley. I recognized Tina's voice. Since we had arrived in camp with Yula and Cameron, Tina had remained out most of the time. A few moments later we saw her emerge from the gully. She walked cautiously toward the trailer and climbed into the small tree behind it. She swayed at Yula and Cameron, then broke off a branch and began to flail the trailer with it. The force she was using almost made me change my mind about releasing the new arrivals, but I felt that I could protect them initially. Tina was terrified of the starter's pistol, and if worst came to worst, I'd fire a shot. After beating the cage several times, Tina returned to the vines behind the shack, where she was joined by Pooh and Bobo. Tina pulled

Yula picking a fig

Bobo confidently toward her and began to groom him. Pooh groomed Tina's back.

Thirty minutes after drinking his doctored orange juice, William was lying a few yards from the trailer in a relaxed attitude, dozing lightly. I decided the time had come. I walked to the trailer and began to unpadlock the door. William sat up immediately and blinked at me, but didn't move. I glanced round to see that Tina had vanished but that Pooh and Bobo were sitting close to one another, watching me intently. I whispered reassurance to Yula and Cameron, lifted off the padlocks and opened the door. They climbed out immediately and walked straight past me to the back of the shack. They had a frighteningly independent air, and Yula was definitely heading away from camp and into the gully. William got up and walked after them. He seemed interested in reaching Yula—to my relief he looked remarkably calm and friendly. Cameron repeatedly glanced over his shoulder at William, and deliberately kept himself between Yula and William. Just before they reached the edge of the gully, William tried to overtake and intercept Yula. Cameron herded her ahead of him and, as he turned to walk back the way they had come, struck out aggressively at William.

William stopped; all appearance of drowsiness vanished. His coat bristled, his shoulders hunched forward and a low, deep pant-hoot began to vibrate in his throat as he grasped a piece of dead wood. Then he flew at Cameron. Yula ran off screaming. Cameron made a brave attempt to defend himself, and after a few seconds Yula ran to his aid and began biting William. Suddenly there was the sound of a branch cracking above the fighting chimps and Tina came down to enter the fight on William's side.

Yula and Cameron together were proving no match for William. If Tina joined in I feared they would be seriously beaten. I raised the alarm pistol and fired twice. Tina screamed and vanished into the gully. William stopped. Cameron and Yula reassured one another with embraces. The screaming was deafening. Pooh and Bobo were right beside me and also started shrieking when the excitement became too much.

William initiated round two by charging at Yula and dragging her down the slope of the gully by her foot. I fired again. He let go of her and charged on into the gully. Cameron ran to his ruffled, leaf-strewn sister and embraced her. Together they hurried back up the slope. I tried to speak to them, but they were still screaming and ignored me. However, when Pooh and Bobo approached them, Cameron hugged Bobo, and then all of a sudden they were all four hugging each other at the base of a small tree. William came strutting up out of the valley just as the greetings between Pooh and Bobo, Yula and Cameron were being concluded. I was very afraid that he would hound the newcomers out of camp if he continued the fight, so I intervened.

William was panting from the exertion, and stopped his advance when I gripped a handful of hair on his back, buried my face in his neck and spoke to him in the most soothing tones I could muster. I embraced and kissed him and tried to calm him down. William allowed me to cosset him till he had recovered his breath, then walked calmly to where Cameron sat by the edge of the shack. Cameron politely stepped out of his path and sat down a few yards away. William looked at him as if satisfied that this strange male was at last learning a few manners, then sat down. The drug I had given him seemed to take over again—William stretched out and slept in the shade. Cameron, Yula, Bobo and Pooh climbed onto the roof of the shack and played for two straight hours. Every once in a while William would open an eye and watch them for a second before continuing his slumbers. I felt exhausted but lightheaded with relief as I watched Yula and Cameron play so confidently with Pooh and Bobo—as if they had known each other for years.

While William slept I gave the four younger chimps some fruit and as much fruit juice as they could drink. Cameron, Bobo and Pooh all fed heartily on the mangoes I'd brought to camp, but Yula didn't like them. I gave her the last two grapefruit. Despite all the screaming and thumping that had taken place during the fight, neither Yula nor Cameron was really hurt: Yula had one or two grazes and Cameron a superficial bite on his foot.

When William woke up he began to make tentative play invitations to Yula. She avoided him at first, still very much afraid, but he persisted and followed her around. Finally, when he lay down and began making comical grabbing gestures toward her, she stopped, looked at him and allowed him to touch her. He made a few grooming motions on her coat, then inspected her behind carefully. The introduction over, they began to play. Yula rapidly gained confidence and seemed to forget completely that her playfellow was the same tyrannical chimp who had stamped on her and dragged her by one foot halfway down the gully.

We had prepared low hammocks for Yula and Cameron in one corner of camp, similar to those in the sleeping quarters at Abuko. Réné and Julian cut a pile of fresh leaves for each hammock, but Yula and Cameron refused even to try them.

While the other chimps were making nests, the two new arrivals headed down into the gloom of the gully. I followed closely, anxious to make them return to camp. I probably would not have succeeded if Yula had not permitted me to lift her and carry her back. Cameron followed closely. Finally they made ground nests by pulling the dry grass down around them and settled in these. I hated the idea of them sleeping on the ground, but had to be grateful that they were at least in camp. I knew that if I tried to insist on hammocks they would probably vanish into the gully again.

The next day we stayed in camp and I did all I could to win the confidence of these strangers. Yula fortunately accepted me almost immediately and lapped up the affection I offered her. By the end of the day I believed that she would trust me in an emergency and run *to* rather than away from me into the unknown. Cameron was friendly yet quite independent, but I felt that he would stay close to his sister at all costs and I could initially control his movements through her.

Those first few days were a nightmare for me. I had to keep continual watch on Yula and Cameron in case they wandered off and got lost. To add to the problem, Yula refused to eat anything strange and I had none of the familiar fruits in camp. Finally I was opening cans of fruit and mixing wild fruits with sugar. Cameron was much easier—he was willing to try new foods and within a few days was eating many of the wild fruits that Pooh and Bobo fed on around camp.

Yula came to idolize William and totally lost her fear of him. She was far more confiding and cheeky toward him than any of the other chimps dared to be, and he in turn found her amusing and good company. It was months since I had seen him so playful and amiable. He tended to be possessive about her, but Yula, in a very feminine fashion, was able to get William to accept almost everything she did.

Cameron became uneasy whenever Yula and William were playing. I was never sure whether he felt jealous of the attention Yula gave William or was afraid that William might hurt her. For several days he tried to intervene and pull Yula away. Yula's influence over William made him relaxed and happy, and he was remarkably tolerant of Cameron's tiresome intrusions into his games. This, in turn, helped Cameron to calm down, and during the evening of the second day the two touched each other. William reached out and held Cameron's foot, and Cameron did not try to pull away. William groomed Cameron briefly, and after this first friendly contact they really accepted one another. Cameron was well aware that William was the dominant chimp in camp, but he did not seem to know how to express this knowledge. I soon realized that Yula and Cameron did not understand all the subtle significance of chimpanzee communication. Even I could comprehend the language better than they, but then I don't really know why I expected anything else, for they had had little contact with other chimps.

Yula's impertinence was partly due to the fact that she was unable to understand the signs and gestures William made when he was asserting his authority. In her innocence she showed no fear or nervousness, and her undiluted confidence initially entranced William. From Cameron, though, he found the same ignorance irritating. When William marched over to Cameron and sat squarely in front of him, it was a clear indication to us all that he wished to be groomed. Cameron had no idea what William was demanding, and would ignore him or try to play—and then could not understand it and got upset when William became irritable.

For three days Yula could do no wrong in William's eyes; then on the fourth morning he came into camp in a bad mood. He strode in, slapping down his feet, with an expression on his face that made Pooh and Bobo hurry out of his way and sit behind the shack. Yula could not read the signs. She sat directly in William's path and slapped him playfully on the shoulder as he passed. William ignored her and strutted on. Yula followed him, pulling at his feet. For a second I thought William might be persuaded to play, but then he snatched at Yula's hand—a clear sign that she was annoying him. Yula gave a startled scream and threw a handful of dirt at him. He spun round and gave her a good beating. Cameron ran to her rescue, but William rushed to a tree. He drummed out the remains of his irritation against the trunk and then calmed down.

I had to be careful in situations like these. I wanted to comfort Yula and Cameron, but not to the extent that I would seem to be favoring them in William's eyes—this would probably make him even more bloody-minded. Fortunately William had learned long ago that if anyone was hurt, I helped them, whether it was Tina, Pooh, Bobo or William himself. On the other hand, if, by not caring for the others, I gave the impression that I was on William's side, he would be likely to repeat the aggression. I had to try to be totally fair—and fair by chimp standards, not my own.

For the first three days I did not move out of camp, not even to go to the stream to wash. I wanted Yula and Cameron to feel settled and to get to know the other chimps and me. On the afternoon of the fourth day I decided I could risk walking to the stream in the hope that Yula and Cameron would follow. As I entered the gully it gave me a tremendous sense of elation to turn around and see five chimps walking in a group behind me. I wanted to run and whoop with joy to release my mounting excitement, but I dared not in case I shattered the still-brittle serenity of the scene.

Yula and Cameron followed well, looking about continually. How high the trees must have seemed, but the new chimps appeared to be comfortable and secure, not apprehensive of this sudden lack of restriction. They were learning to make their own decisions and were choosing to do all the right things to keep themselves safe.

Yula and Cameron seemed very impressed by the stream. Cameron stood and gazed at the water but kept well clear of it. Yula crossed confidently on the large boulders the chimps used as steppingstones and sat down on the last one, staring intently past her own reflection at the small fish that swam slowly in and out of the rocks. She began to shake her hands at them, then leaned forward

Yula drinks from the stream near camp.

and tried to grab them. She was concentrating so on the fish that she seemed to have forgotten the water, and jerked her hand back sharply when it hit the surface. Later Yula—who I'd previously thought of as lacking initiative—picked up a small twig and used it to poke at the fish.

On their first afternoon at the stream Yula and Cameron followed the example of the other chimps and myself and drank from the edge of the water. Yula seemed excited, playful and happy, and rolled her way over to William. William was lying on his back, looking relaxed, and he responded to Yula's high-spirited approach by playing with her. Bobo, Pooh and Cameron also played for a while; then to my delight I saw Cameron join them in a meal of young mandico leaves and apparently enjoy the new food. Everyone looked so peacefully content that I began to undress for the bath I'd been looking forward to.

Suddenly there was a familiar panting on the slope behind me. It was Tina. She had been away ever since I had fired the shots during the fight. I quickly paddled back to my clothes and took out the pistol, which I then hid in my towel. All the tension of the past few days returned. I was sure there was going to be another fight. Would I succeed in keeping Yula and Cameron near me, or would they run off into the valley?

I waited. William looked up at Tina, but Yula was a demanding partner, so he continued to play with her. Tina stood watching them, then came hurrying down the slope, but in a crouched, submissive gait rather than the drawn-up, swaying step that indicated aggression. She hurried directly to Cameron, panting her respect so nervously that he could not fail to see she meant no harm. He grinned and held out his hand. Tina ignored it, placed her own hand behind his head and drew his face into her neck. She then buried her face in his neck, her mouth open, and for a full minute Cameron and Tina remained like this. It was one of the most intense and touching greetings I had ever witnessed. I sat down heavily and almost cried with sheer relief. Tina then groomed Cameron for several minutes, and by the end of the afternoon they were completely at ease with one another. Yula was surprisingly wary of Tina, and avoided her each time Tina attempted to approach. At one stage I thought Yula was going to provoke her into a fit of temper, but Tina finally gave up and returned to Pooh, Cameron and Bobo, leaving Yula to William.

Within a week Yula and Cameron were accepted members of the group. Cameron even then held slight reservations about William, and Yula felt a little uncomfortable with Tina, but we went for walks in search of food together. The new arrivals still had exceptionally soft feet from captivity, and it took several weeks before the soles were toughened enough to make walking on the rocky plateau comfortable. I chose our walks with this in mind, and we remained most of the time on the leafy floor of the valley.

Yula quickly chose me as her guardian angel in camp, and whenever she got into trouble with one of the others, she'd fly to me for comfort. Cameron also came to rely on me in difficulties, but he was exceptionally independent and did not seem to need me as an emotional prop. Cameron became very fond of Tina. At first he unintentionally annoyed her to the point where she attacked him, but he quickly learned the code of behavior adhered to by the group and adapted his own behavior to conform. Though he was still extremely attached to Yula, he no longer seemed to feel the fierce

Yula and Pooh playing

protectiveness toward her that he had shown at first. They became individuals within the group rather than a constant pair.

Cameron was in many ways a much less complicated character than Yula. He was a big, brawny chimp for his age, with an open-looking face and a courageous, direct manner to match. He soon caught on that if he ate what the other chimps ate during the day, he did not have to remain hungry till the evening, when I gave them an uninteresting dish of rice and a cooked sauce of some sort. He was now much more willing to experiment with new foods, and needed scarcely any persuasion from me to try something if he saw the other chimps feeding heartily. He had a liking for leaves, and when the kenno buds appeared he climbed readily with the other chimps and fed on the young leaves with much appreciation.

Yula sat at the bottom of the tree and watched. She was a headache to feed and her loss of weight worried me. Even in camp it was difficult to find something she liked. I got the disturbing impression that she would starve quietly rather than eat anything she did not know. Her attachment to me began to help in this respect, because, to please me, she was prepared to eat things I put in her mouth. Netto pods are sweet and palatable by any standards—all the chimps, Cameron included, loved them, as did the humans in camp. I

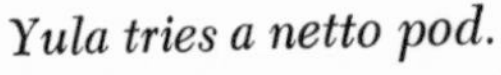

Yula tries a netto pod.

felt that if Yula would only taste them she'd like them too.

I would sit her on my knee, cuddle her, laugh with her and pop a piece of netto pod into her mouth. If she immediately spat it out, I'd whimper, open her mouth and pop it back.

If she even as much as kept it in her mouth for a second, I'd embrace her and praise her ecstatically—something she really loved. It took me a long time and a lot of patience to get just one pod down Yula. It was a delicate process. I had to be insistent yet almost playful at the same time, for otherwise she would close up completely and refuse to do anything. My endeavors were rewarded the following morning. I woke to find Yula sitting beneath the netto tree, picking up all the small pieces of pod as the other chimps dropped them and eating them with an appetite.

That afternoon I climbed up the ladder to the platform in the netto tree carrying Yula. Julian passed me a bamboo pole, and with it I was able to retrieve a bunch of pods. I made such a fuss food-grunting about my prize that Bobo came over to see what was so special about this particular bunch. Ignoring both the chimps, I began to eat, making sounds as if the pods were the greatest delicacy I'd ever tasted. Then I grudgingly gave Bobo a piece. Yula watched me intently until, almost as an afterthought, I offered her a scrap. Yula was by this stage so enthralled by the whole process that she immediately stuffed the piece into her mouth and ate it. She did not beg for more, so I continued eating and a few moments later passed her a slightly larger piece. This she took just as enthusiastically. Then I stretched out on the platform, pretending to be full, and left the bundle of pods a foot or so away from me. As I had hoped, Yula picked up the pods and began to open them and feed. She finished off the whole bundle. I used the bamboo pole to pick another bunch and openly handed them to her. To my relief she accepted them and started eating.

The next morning she was feeding alone, along with the other chimps. I had to go through this process with almost every new food as it came into season. Slowly Yula was learning, and the more she learned the more open-minded she became about accepting new foods and new situations.

About two weeks after Yula and Cameron were released we heard the sound of a vehicle approaching. It was not my father's car but a yellow Land Rover that came bouncing across the plateau toward us. Inside was Raphaella.

William reached the plateau before me, excited by the prospect of a visitor. With the attack of Charlene still fresh in mind, my greeting to Raphaella was, "Be careful of William." We stopped to watch what would happen next. I had my hand in my pocket, my fingers tightly wrapped around the starter's pistol. Raphaella leaned out of the window and spoke quietly in warm welcome to William. William's hair gradually flattened, and Raphaella stepped out of the Land Rover. She held out her arms, and with very evident and real affection embraced William, pulling his head onto her shoulder and panting and talking to him alternately. William embraced her, then patted her back gently several times. As was the case with Yula, Raphaella's confidence and open affection had disarmed William completely. Also, William remembered Raphaella as a person who knew the rules and who had an unlimited depth of affection and understanding—she was not an ordinary person.

I watched her with growing admiration. She had handled William perfectly. Before doing anything else she gave him a pile of mangoes to keep him occupied, then drove into camp. Bobo stood by with Pooh; Yula and Cameron were afraid of the Land Rover and had crept round to the back of the shack. Suddenly Bobo realized who was in the car. With a scream of joy he scrambled into Raphaella's arms. She hugged him tightly, biting her bottom lip to stem the emotion she felt at holding him again. Tears squeezed through her tightly closed eyes, and it was a moment or two before she could speak. Her words were for Bobo alone. He listened intently and kept kissing her face, a human gesture of affection he still remembered.

Then Raphaella slid out of the seat and embraced me almost as tightly as she had Bobo. All my British cool left me and I hugged her back. We laughed at each other, self-conscious about the happiness we felt.

Raphaella was to stay for a month. Apart from being a terrific help, she was great company. William respected her, but she was not immune from attack if he was in an excitable mood. One morning after breakfast we stood on the plateau to watch a storm ap-

Yula and Raphaella

proach. Then Raphaella went off to the lavatory and I returned to the house. I was helping Réné refuel the fridge when I heard a loud, authoritative shout from Raphaella, followed by two pistol shots. William went careering past the window with Raphaella in full pursuit—one hand holding up her trousers.

I tore out of the shack in time to see William stop on the edge of the gully, pick up a sizable rock and fling it at his pursuer. The rock missed Raphaella's head by about six inches. Blaspheming in Italian, she bent down, picked up a rock, stepped out of her trousers and went hurtling down the gully after William. She threw her rock and it landed close enough to William for him to give a squeak of surprised alarm before vanishing into the vegetation. Raphaella stopped and, breathing hard, climbed back up the slope to camp.

It was only then that I saw the red gash on her knee. I went to her and handed her her trousers. She was white and trembled slightly, but giving me a wry smile, she put the trousers on carefully, then rolled the right leg up above the knee and inspected the wound. It was not a bite but a gash where William had pushed her over a rock. We walked slowly to the shack, and Raphaella sat down on the bed. Réné brought some warm water, and I washed the dirt and blood away from the wound. It was a deep, clean cut about an inch and a half long. Raphaella looked at it closely, then asked if I had any suturing equipment. I had a little box full of assorted needles and gut, but no anesthetic. Undeterred, she took the needles.

I watched with mingled horror and fascination as she put four stitches in her own knee. Réné and Julian couldn't believe their eyes, and after the first stitch had been efficiently tied, Julian turned away and stayed in the kitchen till all was over. During the afternoon William returned. Raphaella went straight over and sat grooming him or playing with him for a long time—talking to him in cascading Italian. Both of them seemed to have forgiven and forgotten their disagreement earlier in the day.

27. A STEP TOWARD THE FUTURE

Yula and Cameron were making good progress. It was perhaps not surprising that they were able to recognize fruit trees easily, after they had eaten from them once, but I was pleased that they also knew the shapes of kenno and other trees which provided no easily distinguishable fruits but merely edible leaves or bark. They would begin to food-grunt in anticipation long before they reached kenno or young kapok trees, and climbed readily to feed. Their climbing techniques had improved till they were almost as agile as the others. Yula once misjudged the strength of a branch, which broke and fell to the ground with her. She was winded and shocked by the fall and suffered a couple of bruises, but to my relief it did not make her afraid to climb high in a tree—merely more careful.

The soles of their feet were also beginning to harden, and they were able to walk longer distances each day. At first Yula often whimpered for me to carry her, but she soon became more confident and, instead of asking to be carried, formed the habit of walking ahead of me to keep me at her pace. If I overtook her, she began to scream. I realized that this was Yula's own method of conquering her fear of being left behind, so I let her have her way and lead. Cameron often lagged behind, not because he was tired but because he had not grown up in the routine of following someone and did not seem to feel the need to stay close to the security that a human represented. Once or twice I had to go back for him or wait for him to catch up, for I feared we might lose him somewhere and did not trust his ability to find his own way back to camp. Cameron still had a lot to learn before I would consider him self-sufficient enough to go it alone if he chose.

After three weeks Cameron began to abandon the low hammocks and search for a higher sleeping site. Of his own volition he then chose to sleep on the shack roof. He would collect his bundle of fresh-cut leaves from the hammock and carry them up onto the tarpaulin. There he would spend several minutes arranging them into a nest before finally lying down to sleep. It took Yula much longer to forsake her hammock for a loftier sleeping platform, but when she did so she went straight into an old nest one of the others had made.

Four months after their arrival I was ill enough for it to be essential to visit a doctor in The Gambia. I hated to leave the chimps when they were still not entirely acclimatized, but Nigel was able to come up to take my place, so I knew they would be well looked after. I was called back by a telegram from Nigel informing me that Cameron had left camp with Tina and had not returned. By the time I got back to the camp Cameron had been missing for nine days. I searched with Nigel for another three days, but we found no trace of him. Yula was well integrated into the group and didn't seem to miss her brother.

Nigel told me that Cameron had become very close to Tina and often walked near her. They fed and groomed together more than with the others. William had often intervened, and although the two oldest males did not actually fight, their relationship was a tense one. Cameron was never completely submissive to William, and this rather unnerved the elder chimp. He became overaggressive toward the younger ones and frequently attacked them.

One afternoon when Nigel returned from a walk, Tina and Cameron, who had been last in the line of chimps following Nigel, remained in the valley instead of climbing the slope to camp. But since they were barely two hundred yards from camp, Nigel did not worry. William came with Nigel almost into the camp, then went back into the valley. Nigel presumed he had gone to join Cameron and Tina. Two hours later William returned to camp alone. Nigel began to get slightly anxious about Cameron and went into the valley to check that all was well. It was unusual for him to stay away so long.

The valley seemed deserted. Nigel could find no sign of either Tina or Cameron and finally had to return to camp, as it was getting dark. The following morning there was still no sign of the two missing chimps, so Réné, Julian and Nigel searched the valley all day, calling out to them. Each day they searched farther afield. On the third day Tina returned to camp, but Cameron was not with her.

Finally Nigel and I gave up searching. Nigel returned to The Gambia, so once more there was only Julian, Réné and I in camp. I missed Cameron, and on all the walks I kept my eye open for traces of him. It was difficult to imagine what had happened to him, but I suspected that he had followed Tina on one of her long excursions. Perhaps he had got tired and stopped, or perhaps Tina had left while Cameron was sleeping, but if he had been separated from her miles from camp, I was sure he would be lost. I felt fairly certain that he could feed himself. It was the rainy season and there was plenty of obvious fruit around. If he could not find a water hole, there would be rain puddles to quench his thirst. But he still lacked an awareness of the bush. I wasn't sure that he could cope with the dangers and the loneliness of life in Niokolo without supervision.

One possibility was that he might be with a group of wild chimps. He was still young, so they might have accepted him if he had wanted to join them. If Cameron is with wild chimps, I thought, there is a good chance that sometime this rainy season they will visit the valley. Cameron may pay the camp a visit if he finds himself on familiar ground. There was little I could do but wait and hope.

Though Yula never seemed actually to look for Cameron, she became much less adventurous and still more emotionally dependent on me. In fact, I became the center of her life. It was relatively easy to teach her to eat the new foods coming into season if it was just a matter of climbing and eating, but it took me a long time to show her how to crack open a baobab shell or an afzalia pod. As far as Yula was concerned, the fruit of the baobab was securely locked in a thick, hard shell which she couldn't open, so there was no point in climbing the tree to get it. The same was true of the afzalia pods. Yula just didn't bother. If I put her in an afzalia tree she would sit on a branch and daydream or sleep. I thought that after watching Pooh and Bobo climb, pick their pods and bring them down, Yula would get the idea and fetch her own pods, but after almost three weeks she still refused to try. Yula was not stupid; in certain ways she was, indeed, exceptionally bright; but when it came to picking baobab or an afzalia pod, she seemed to have a mental block.

One day we came to an afzalia tree that I could easily climb. I decided that somehow I would get Yula to follow all the steps of climbing, picking and then opening, or at least seeking help to open, her pod. I climbed the tree with her. We sat on a branch together and I pointed at a pod and held out my hand. Pointing and holding out the palm of my hand were signs Yula understood. I was clearly asking her for the pod. Yula merely looked at me. I pointed, held out my hand again and gently urged her with words as well: "Give me the pod, Yula; give me the pod!" I repeated this over and over again and made chimp food-grunts. There was still no response from her. She merely watched my mouth as I spoke.

Bobo climbed the tree at that moment, came close to us and picked three pods. I praised him loudly for Yula's benefit. Bobo looked round at me in surprise and acknowledged the unusual praise by panting into my hair briefly on his way down the tree. Again I tried asking Yula, hoping Bobo's example might have made things easier for her to understand. Still I saw no gleam of comprehension in her eyes.

I picked up her hand and wrapped her fingers round a pod, my voice enthusiastic to indicate that she was doing the right thing. Then I held out my hand again, speaking to her all the time. She held the fruit briefly, then put her hand in mine, almost as if she were miming the action of giving me something. I put her hand round the pod again and placed her other hand on the small branch from which the pod hung. Then I placed my own hands over each of hers and, still talking encouragingly, tightened my grip around her hand on the pod and pulled it away from the branch. The short, tough stem by which the pod was attached tore away from the branch and Yula found herself holding a pod. I threw my arms around her and hugged her tightly in exaggerated praise. She was surprised at my excitement but obviously pleased. I held out my hand and Yula placed the pod in my palm. I hugged her again, food-grunting like mad. With Yula on my back I hurried down the tree, went immediately to a flat rock, hammered the pod open and handed her the open pod full of flat, crunchy purple seeds.

Yula food-grunted enthusiastically and ate. While she fed I

climbed the tree, broke off a branch with six pods dangling from it and brought this down to the ground. When Yula had finished she hurried over to where I sat with the branch. I pointed to a pod and held out my hand. Yula briefly touched a pod and then mimed giving it to me. I shook my head, my voice slightly stern as I said, "No, give me the pod, Yula!" Again she touched the pod and looked away. I put her hand around the pod again, placed my hand over hers and pulled. The pod came away. I praised Yula again. I held my hand out; she gave me the pod and waited impatiently while I hammered it open for her as quickly as I could. With each repetition of this process Yula became more self-sufficient, until for the fourth pod she needed almost no help. I pointed and held out my hand. Hesitantly she placed her hand on the pod and pulled it off by herself. I merely held the branch. I praised her profusely, hugging and kissing her; then, as soon as she held out the pod for me to take, I hammered it open and handed it back. We panted and food-grunted at each other, and Yula climbed into my lap to eat her pod.

When she had finished I sat her down on the ground beside me. I picked a pod from the branch, walked the few yards to the tree and placed it in a fork. Yula was watching me. I walked back and sat down. Pointing at the pod in the fork, I asked her to get it for me. To my delight she got up almost immediately, went to the tree and came back food-grunting with the pod in her mouth. She handed it to me immediately and I opened it for her.

While she ate I took the branch, with its one remaining pod, to the tree and wedged it well into a fork. Yula was watching me. When she had finished eating I said nothing, hoping she would go and pick the last pod without any encouragement from me. She didn't, but sat next to me, rocking slightly. I knew if she didn't hurry Pooh and Bobo would take the pod, so after a minute I turned to Yula, pointed to the tree and held out my hand. She went over right away, brought the whole branch back with her and gave it to me. For a second I wasn't sure what to do. I wanted her to pick the pod, and wondered whether I would confuse her by insisting she do so after she had so confidently brought the branch to me. Finally I decided that it would confuse her more if I made exceptions, and asked again for the pod. She hesitated slightly, but after I'd pointed to the pod a couple of times, she pulled it off the branch.

The following day we went back to the same site. I carried Yula over to the tree and put her up into it, pointing at the pods and holding out my hand. I was bitterly disappointed at her lack of response. I climbed with her till we were close to some pods, and again went through the pointing routine. She ignored me until I ordered her to give me a pod. Then she looked up at me sharply, slightly worried that I seemed to be getting annoyed, and came over to hug me. I softened my voice slightly and asked again. Yula reached out and pulled a pod toward us. She held it, staring up at me. I encouraged her and she came out of her trance, picked the pod and again was rewarded by ecstatic praise.

The second time I sent her up, I remained on the ground. She climbed to within reach of the pods, then sat and stared at the ground in her trancelike way. My coaxing became orders, but still she ignored me and just sat. I picked up a small stick and chucked it at her. She started and began to whimper. "Pick a pod, Yula!" I commanded. She felt unsure and worried that I, the idol of her life, was getting angry with her, and she began to climb down. "No," I barked. She stopped. Immediately I softened my voice, and she ceased whimpering. I pointed at the pod again, gradually pretending to get angry. Yula hurried to it and picked it quickly. I praised her, and she began to scream with excitement as she hurried down the tree with the pod in her mouth.

For a week I had to coax or bully Yula into climbing for the pods, but finally she was doing so without being asked. She had to do a lot more work for her food than Pooh and Bobo, though, for she only picked one pod at a time, whereas the other two had learned long before that it was much easier to pick as many as they could carry while they were in the tree.

This phase did not last long, however. Once Yula had reached the stage of racing Pooh and Bobo up a tree and picking a pod as a matter of course, she also started to collect them. At first it was a case of a second pod that happened to be right under her nose as she climbed down; then she began to search for two pods before descending. The rewards of collecting were self-evident once she began. In time she picked three or four pods before descending to have them opened. Every month Yula becomes more inventive and receptive to new ideas. Soon, I am sure, she will learn to use a rock

Yula picks and opens a baobab fruit.

to open her pods. Even if she doesn't, however, by helping her to make pods an important food item, she will be much more likely to persevere and to open them with her teeth when she has her permanent canines.

One night we found a snake coiled round a piece of bamboo just above Julian's bed. It looked like a type of cobra and was about four feet long. Julian procured a machete, or "coup coup" as he called it, and while I shone the flashlight on the snake, he inched into the doorway and in one quick slash severed its head. The body writhed round and round on the bamboo pole. Julian found a stick, carried the snake's still-writhing length onto the plateau and flung it away. The disturbance had brought Yula out of bed. I put her back on the ladder and she appeared to climb up it, but ten minutes later she was back at my window. I ignored her for a while, then took her back to the platform. When I went out about an hour later, Yula was sleeping on some dry leaves beneath the window by my bed. I scolded her, put her back on the ladder and waited till she reluctantly climbed up and settled. It worried me considerably that she felt at ease enough in the dark to sleep on the ground.

During the next few weeks I regularly checked all around before going to bed, and not infrequently discovered Yula sleeping on the ground, even though she had seemed to settle on the platform earlier. How could I make her understand that it was dangerous to sleep on the ground—especially when she slept so heavily? She was vulnerable to scorpions, snakes or, even worse, prowling hyenas and other predators.

I remembered the pods and how I'd had to show Yula the way step by step. Months ago I had found a dry buffalo skull on the plateau. All the chimps had shown extreme nervousness and fear of it; in fact I'd even hung it on the Land Rover to keep William from wrecking the car. One night when I found Yula on the ground, I did not wake her. Instead I went to the shack and asked Julian to help me. I dressed him up in a striped blanket and gave him the large buffalo skull to hold in front of his face. He put his flashlight beneath his chin so that the skull was illuminated. Then I handed him a whistle to blow, so he would not have to speak to make a noise. I told him to creep toward Yula and, when still about ten yards from her, to begin to blow the whistle gently to wake her up. I wanted to frighten her but not to give her a heart attack. Then, when she was awake, he was to run at her as if he were going to catch or attack her.

I followed Julian out. Soon I heard the whistle blowing softly but rhythmically, and then Yula gave a terrified scream. I called her name, ran toward her, swung her into my arms and sped to the ladder. Julian chased us there. I feigned desperate fear and hugged Yula close while Julian danced around at the base of the ladder. Yula, Bobo and Pooh whaaed and barked at the monster, who then hopped away into the darkness of the plateau. I settled Yula into her nest, then, still acting nervous, hurried off to my bed.

Two weeks later Yula tried sleeping on the ground again and I arranged for the monster to return. After her second lesson Yula learned to remain on the platform or in a nest once it got dark.

One evening as I was preparing to go and wash I heard wild chimps calling just down by the stream. The trees above the stream were full of chimps feeding on pods. I hid and watched them. Most of the time I could only see an arm or a patch of black hair through the foliage. There was much food-grunting going on, and several times I heard young chimps whimper or scream. To my amazement the chimps nested in the trees above the stream—either oblivious or indifferent to their proximity to camp. I waited till all the nesting sounds stopped, then returned to camp. William was missing.

I woke early the following morning, intending to take a position from where I could easily watch the chimps getting up. It was still pitch dark as Julian and I walked into the gully by the beams of our flashlights. From our vantage point we heard rather than saw the wild chimps rising. There were choruses of pant-hoots and frequent drumming before the group began to move downstream. Suddenly I stiffened, for I clearly heard William make the submissive coughs he used when nervously approaching someone he considered dominant. There was another burst of submissive coughs, this time rising in pitch to end in a short scream. Again I could swear it was William's voice.

I waited and then followed carefully at a distance. The chimps were moving rapidly, and having always to keep to cover, I soon got left behind. William returned that evening and flopped down on the anthill. He looked tired and, whether he had been with wild chimps or not, he appeared to have just come back from a long walk.

One Sunday morning, three months after Cameron's disappearance, I was sitting in the shack writing when I thought I heard Yula begin to scream. I got up and hurried to the door. As I reached it I suddenly realized that the chimp screaming wasn't Yula: it was a very similar voice, that of her brother Cameron. With a mounting sense of excitement I ran out to the plateau. William, Pooh, Yula and Bobo were sitting in a group staring out toward the sounds, which came from the top of the gentle slope that rose from the east side of the plateau. The screams were not those of a chimp being attacked, but the same long, drawn-out shrieking William used to make as he approached camp after spending days away with Tina.

I began to call Cameron's name loudly over and over again as I ran out across the plateau. I was convinced that he had suddenly found himself in a familiar area and was on his way to camp. The screaming continued, and I kept yelling. Bobo and Yula, alarmed at my haste, were running after me, whimpering. Yula finally threw a temper tantrum because I refused to wait. I hoped her voice would attract Cameron. I got close to the screaming, but the grass was high and the vegetation thick. I repeatedly shouted, expecting any moment to see Cameron emerge and walk toward me. Suddenly the screaming stopped. I continued to call, but there was not a sound.

At the top of the slope the land suddenly drops into a large, round basin. Along its edge I found a chimp track, made, however, not by one but by a small group of chimps. Just opposite where I'd heard the scream the trail turned sharply and led down into the dip. We searched for the rest of the day without success, and for the next few days I walked the chimps in that direction, but found no traces of what might have been Cameron. I trusted my ears and my ability to distinguish the voices of my chimps. They are as different as human voices. If I had been mistaken, if it had been a wild chimp screaming, he would not have allowed me to run so far toward him. I was very close to the screaming before it suddenly stopped. Yet it was even more difficult to understand why, if the chimp *was* Cameron, he had disappeared so suddenly without showing himself or coming into camp to see the other chimps. If he was with a group, perhaps they had run away when they heard my voice, and he, torn between camp and new friends, had run after them. The day left me disappointed and puzzled. I had been so sure I was about to see Cameron again.

IT WAS NOT TOO LONG before I was compensated for this disappointment. One afternoon I had been working with Nigel on the Land Rover. At about five o'clock there was a burst of screaming from the gully, as if two of the chimps were having a squabble. The screams were short-lived, so I did not worry unduly. Then I heard Réné calling my name—his voice high with urgent excitement as he careered toward us at top speed. His face was split in an ecstatic smile.

"Come quickly, oh come quickly! Tina has just come into camp with a baby!"

Tina termiting with Tilly

Tilly

I screamed with joy, then flew back across the plateau. Tina was sitting on the ground near the shack. Pooh and Bobo were staring curiously at her stomach and William was peering over her shoulder, but Tina kept moving round protectively, cuddling her newborn baby. As we approached, she climbed into a low tree. She had to use both her hands, but she pinned the tiny infant to her stomach by crossing a leg over its back. She sat on a low branch, and I was able to see a tiny red fist clutching the hair on her side.

The baby was covered in a soft-looking dark down, his head set in a silky cap of black hair. His face was dark and screwed-up. His mouth was a bright red line, and his eyes, when they opened, were unfocused and a light beige-brown. Already his small ears stuck out like his father William's. I felt delirious with excitement. I asked Réné and Julian to bring out all the bread they had baked that day and some rice. I gave the chimps rice, and when they were all occupied I handed Tina a large loaf of bread, which she loved. As she leaned forward to take the loaf, I saw the baby was definitely a male. The light was poor but we took pictures anyway, then I sat and watched Tina.

Any fears that I had harbored about Tina being an inadequate mother soon vanished; she was very careful about supporting the baby each time she moved, and appeared tender and proud of him. While she was eating her loaf the baby began to move his head jerkily from side to side on her belly, and she considerately hitched him up higher so that he could nuzzle the area around her nipple. After a few seconds he found what he was looking for, fed and then fell asleep again.

When Tina finished her bread she climbed to the ground, always keeping either her hand or her thigh against the baby's back. As she reached the ground, the baby lost his grip on her coat with both his feet and I heard a tiny, squeaky whimper. Tina stopped, looked down anxiously at him and held him closer, then walked away on three limbs—her right hand continually supporting her baby. She climbed into a nearby tree and, despite the handicap of being able to use only three limbs, made a lush nest and settled down. After their initial curiosity the other chimps paid Tina little attention that evening.

Réné and Julian had made a special supper, and while we ate we thought of names for the new addition to our family. Finally we decided on Tilly, a combination of the names of his parents, Tina and Willie.

Tina kept to the valley and remained with the other chimps most of the time. No one was allowed to touch the baby, but on his third day in camp I saw William's hand move closer and closer to its small red foot as he groomed Tina's back. Finally he gently lifted the minute toes with his thick, calloused forefinger and stared at them, then quickly continued to groom Tina.

EPILOGUE

OF THE ORIGINAL TRIO that I left behind in Niokolo Koba five years ago, only Tina's progress has been followed. Albert's whereabouts still remains a mystery. Cheetah too has vanished; but he did prove that he was able to survive his first year, and it is not unreasonable to hope that he continues to live somewhere in the park.

As I look around camp today, I cannot help feeling enormous satisfaction at what I see. We are more organized now as a rehabilitation center than we were in those idealistic, pioneering days. The shack has made way for a larger, more solid building in which we are able to cook and keep our supplies and equipment out of the reach of the chimps. This has eliminated many frustrations for both the chimps and ourselves; and still more important, I no longer have to battle with an ever more powerful William to protect my possessions.

Tilly's birth proved the success of one vitally important aspect of the project: a chimpanzee, having spent years of her life in captivity, has not only been reintroduced to the wild, but has mated and given birth there.

Tilly is now six months old, and a strong, healthy infant. Already he is able to reach up from his mother's protective embrace and taste her food, the wild fruits that will sustain him for the rest of his life.

William, that undernourished scrap of life that we bought so long ago, is now over ten years old and a father. He has broadened into a fine, muscular adolescent and has become the dominant male of his own small group of free chimpanzees. He is entirely independent in every way and spends most of his time in the valley with Tina and Tilly.

Pooh, once the lonely, insecure baby of the group, has grown into a lanky juvenile of seven and a half years. He is sociable and full of confidence, already beginning to imitate William's impressive male displays. He has become the most agile chimp I have ever seen. He still uses these skills to play the clown, but his knowledge of the valley and its different foods is as sound as William's. Beneath the comic expressions and gestures, he is well aware of his environment. The slightest signal from William, Tina or myself, even a mere tensing of the body, immediately makes him stop playing and become alert. He is little short of a genius when it comes to getting into a tree that cannot be climbed conventionally. Several times when I have thought neighboring trees too far away to provide a bridge, Pooh has discovered a branch he could sway, a twig he could reach with his toes and pull toward him or a vantage point from which he could make one of those long, agile leaps that make my heart stop.

Bobo is Pooh's closest companion. He has fulfilled all his early promise and continues to learn quickly. Though perfectly able to feed himself, he does not as yet have the confidence to wander far into the valley alone. He is young and still requires the security of the camp and occasional reassurance from one of us. He will probably continue to do so until he is about seven or eight years old, when like a wild chimp at that age, he will naturally acquire the independence needed to wander out alone.

As to the zoo-born chimps: Cameron is of course no longer a member of our group. On four occasions since he left, a young male chimp has been sighted that could well be him, each time in the vicinity of wild chimps. Unfortunately none of these sightings could be absolutely confirmed, so we continue to hope for more positive information about him.

Yula is scarcely recognizable as the same chimp that arrived in camp. She has grown into a graceful, long-limbed adolescent and is a well-established member of the group. As I watch the chimps feeding in the valley today there remains nothing in Yula's behavior to indicate that she began life here with a much greater handicap than any of the others. She climbs, feeds and interacts as easily and as normally as any chimp in the group. She is fascinated by Tilly and whenever Tina permits it, she grooms the new baby. Later, perhaps, Tina will allow her to hold Tilly for short periods. Not only is Yula gathering the knowledge by which she will survive, but Tina and Tilly are providing her with experience that I hope one day she will use to ensure the survival of her own children. I am proudest of all of Yula's achievements, as she really had to start from scratch. The fact that she has succeeded in adapting herself to the wild so well proves that, even though a chimpanzee is born in captivity, it can still learn to live the life of its wild ancestors.

It now appears that the camp chimps are unlikely ever to become part of a wild chimpanzee community and so it is more realistic to aim toward creating a self-sufficient group of rehabilitated chimps. Our present six members are an ideal foundation and at the time of writing three more confiscated infant chimps are enjoying the sanctuary of Abuko in preparation for the wider, wilder life here at Mount Asserik.

Finally for me there is no longer the sense of loneliness which used to mar the harmony of camp. Raphaella has returned to stay. She has become my partner and together we will continue with the work. My dearest hope is that not only will our present group of chimpanzees continue to prosper and grow into a viable community, but also that our efforts will inspire others to do the same in those remaining parts of Africa where the chimpanzee is allowed to live and reproduce in peace.

INDEX

A Note on the Type

This book was set in Caledonia, a Linotype face designed by W. A. Dwiggins. It belongs to the family of printing types called "modern face" by printers—a term used to mark the change in style of type letters that occurred about 1800. Caledonia borders on the general design of Scotch Modern, but is more freely drawn than that letter.

Composed by Maryland Linotype Composition Company, Inc., Baltimore, Maryland. Printed by The Murray Printing Company, Forge Village, Massachusetts. Bound by American Book–Stratford Press, Inc., Saddle Brook, New Jersey.

Typography, binding design, and maps by
VIRGINIA TAN